STRATEGIZING SOCIETAL TRANSFORMATION

Knowledge, Technologies, and Noonomy

STRATEGIZING SOCIETAL TRANSFORMATION

Knowledge, Technologies, and Noonomy

Vladimir L. Kvint and Sergey D. Bodrunov

First edition published 2023

Apple Academic Press Inc.
1265 Goldenrod Circle, NE,
Palm Bay, FL 32905 USA
760 Laurentian Drive, Unit 19,
Burlington, ON L7N 0A4, CANADA

CRC Press
6000 Broken Sound Parkway NW,
Suite 300, Boca Raton, FL 33487-2742 USA
4 Park Square, Milton Park,
Abingdon, Oxon, OX14 4RN UK

© 2023 by Apple Academic Press, Inc.

Apple Academic Press exclusively co-publishes with CRC Press, an imprint of Taylor & Francis Group, LLC

Reasonable efforts have been made to publish reliable data and information, but the authors, editors, and publisher cannot assume responsibility for the validity of all materials or the consequences of their use. The authors, editors, and publishers have attempted to trace the copyright holders of all material reproduced in this publication and apologize to copyright holders if permission to publish in this form has not been obtained. If any copyright material has not been acknowledged, please write and let us know so we may rectify in any future reprint.

Except as permitted under U.S. Copyright Law, no part of this book may be reprinted, reproduced, transmitted, or utilized in any form by any electronic, mechanical, or other means, now known or hereafter invented, including photocopying, microfilming, and recording, or in any information storage or retrieval system, without written permission from the publishers.

For permission to photocopy or use material electronically from this work, access www.copyright.com or contact the Copyright Clearance Center, Inc. (CCC), 222 Rosewood Drive, Danvers, MA 01923, 978-750-8400. For works that are not available on CCC please contact mpkbookspermissions@tandf.co.uk

Trademark notice: Product or corporate names may be trademarks or registered trademarks and are used only for identification and explanation without intent to infringe.

Library and Archives Canada Cataloguing in Publication

Title: Strategizing societal transformation : knowledge, technologies, and noonomy / Vladimir L. Kvint and Sergey D. Bodrunov.
Names: Kvint, V. L. (Vladimir L'vovich), author. | Bodrunov, S. D. (Sergeĭ Dmitrievich), 1958-author.
Description: First edition. | Includes bibliographical references and index.
Identifiers: Canadiana (print) 20220399638 | Canadiana (ebook) 20220399689 | ISBN 9781774914229 (hardcover) | ISBN 9781774914236 (softcover) | ISBN 9781003362555 (ebook)
Subjects: LCSH: Technology—Sociological aspects.
Classification: LCC HM846 .K85 2023 | DDC 303.48/3—dc23

Library of Congress Cataloging-in-Publication Data

CIP data on file with US Library of Congress

ISBN: 978-1-77491-422-9 (hbk)
ISBN: 978-1-77491-423-6 (pbk)
ISBN: 978-1-00336-255-5 (ebk)

About the Authors

Dr. Vladimir L. Kvint is a leading scholar in the field of economic strategy and developer of the Global Emerging Market Theory and General Theory of Strategy. He is a member of the Bretton Woods Committee and a fellow of the World Academy of Arts and Science. He is also a laureate of the Gold Medal of Nikolay Kondratyev.

Dr. Kvint is the author of *The Global Emerging Market: Strategic Management and Economics* (Routledge 2009) and *Strategy for the Global Market: Theory and Practical Applications* (Routledge 2016).

In 2018, Dr. Kvint was awarded the Lomonosov Prize, the highest honor of Lomonosov Moscow State University, for his scientific work *The Theory of Strategy and the Methodology of Strategizing*. This marks the first time that this most prestigious Russian academic honor was awarded to an economist.

Dr. Kvint is Chair of the Economic and Financial Strategy Department at Lomonosov Moscow State University's Moscow School of Economics and Head of the Center for Strategic Studies at Lomonosov Moscow State University's Institute of Mathematical Research of Complex Systems. He is a chief researcher and member of the Academic Council of the Central Economic-Mathematical Institute of the Russian Academy of Sciences. In addition, Vladimir Kvint is Chair of the Industrial Strategy Department at the National University of Science and Technology "MISIS." He is also a US Fulbright Scholar, Academician - Foreign Member of the Russian Academy of Sciences (lifetime), and Honored Fellow in Higher Education of the Russian Federation.

Vladimir Kvint was a Visiting Professor at Vienna Economic University (1988–1990), Distinguished Visiting Professor at Babson College (1991), and Professor of Management Systems and International Business Strategy at Fordham University' Graduate School of Business (1990–2004). He was also an adjunct professor of Emerging Market Strategy at New York University's Stern School of Business (1995–2000) and Professor of International Business at American University in Washington, DC (2004–2007). In 1992–1993 and 1997–1998 Dr. Kvint was an economic advisor to the President of the United Nations' General Assembly and was

Director of the Emerging Markets Department at Arthur Andersen in New York (1992–1998).

Dr. Vladimir L. Kvint has research and/or state awards from Austria, Albania, Belgium, Bulgaria, Great Britain, Kazakhstan, Kyrgyzstan, Russia, Slovenia, Turkey, Ukraine, Uzbekistan, and the United States. He is Editor-in-Chief of the *Russian Journal of Industrial Economics* and *Strategizing: Theory and Practice,* and Associate Editor of *Economics and Mathematical Methods, Administrative Consulting,* and member of the advisory board of the *International Journal of Emerging Markets.*

Dr. Sergey D. Bodrunov is Corresponding Member of the Russian Academy of Sciences (lifetime). He is a founder and Director of the St. Petersburg Institute for New Industrial Development (1997) named after Sergei Y. Witte. The main subject of the Institute's exploratory activity is a broad range of issues related to economic industrialization and the forecasting of middle-term and long-term institutional, economic, and technological effectiveness.

He is the author of over 700 research papers, including 30 monographs on the issues of information systems development and society informatization, science and technology progress growth, intellectual property, economic history, and the strategy of high technology economic development. He is the author of fundamental papers on the concept of the new industrial society of the second generation and the theory of noonomy. In particular, Dr. Bodrunov's book *Noonomy* (2018) was awarded the World Association for Political Economy (WAPE) prize for an "outstanding contribution to the development of Political Economy in the XXIth century." Courses in noonomy theory are taught in several universities.

Dr. Bodrunov is the President of the Free Economic Society of Russia, one of the oldest civic organizations of Europe and the world (founded by the decree of the Empress Catherine the Great in 1765) and which includes over 300 thousand members. He is also President of the International Union of Economists, which unites the representatives from 48 countries and has General Consultative Status with the Economic and Social Council of the United Nations since 1999.

Since 2019, Dr. Bodrunov (together with the President of the Russian Academy of Sciences Dr. Aleksandr Sergeyev) has been a co-chairman of the Moscow Academic Economic Forum (MAEF)—one of the largest national research and academic platforms.

Extended Abstract of Strategizing Societal Transformation: Knowledge, Technologies, Noonomy

This book constitutes the first joint experience of the scholars who authored the concepts of Strategizing (V.L. Kvint) and Noonomy (S.D. Bodrunov). The idea of combining the authors' perspectives on issues of general civilizational development has a common scientific platform, i.e., the identification of long-term goals and the selection of economic and strategic tools for achieving these goals.

This publication reflects the co-authors' assessments and perspectives on the global trends transforming society in the modern era, on strategic priorities and public development goals driven by these trends, and the strategic paths achieving national interests.

In "A Word to the Reader," the authors explain methodological premises that serve as the foundation for their joint study. For example, the approach realized in the concept of Noonomy allows for discerning very remote horizons of public development and perceiving transitions from one development stage to another. But a vision of the future does not hold value in and of itself. It acquires public significance when it becomes a tool—a technology which allows for using this knowledge. Methods of strategizing serve this very purpose—they convert knowledge about the future into an instrument for development process management.

The concept of noonomy is a perfect fit for the strategic approach for it incorporates foresight on future events that are obscure from the perspective of common sense, as well as the ability to look far beyond the bounds of the current agenda. A strategy should not rely on vox populi or on rather predictable future elements. Inertial progress that is reliant on the extension of currently implemented scenarios is quite possible without special efforts in development strategizing. A real strategy should take us down an obscure path into a still uncharted future.

Chapter 1: Global Developmental Trends presents the authors' stance on foundations that support the vision of strategic prospects. Any strategy, regardless of how long-term it may be, relies on real-life facts and objective processes that determine the development of the society and its various subsystems. The difference between a strategic and a regular approach is that in its analysis of the established course of events the former discerns patterns that underlie future changes. It is these upcoming profound shifts in public development that dictate the formation of a strategy's goals.

In the future, a country's economic competitiveness will be determined by its ability to pioneer the development and application of high technologies and ensure the quality of human capital and the application of human potential as opposed to prioritizing the extraction and sales of natural resources. New economic leaders will be forged from technological powerhouses. Experts suggest that if current development trends persist (and we must consider that crisis processes allow for certain adjustments) in 2020–25, the sixth technological mode will serve as the foundation for the advance of a new scientific, technical, technological, and industrial revolution.

This revolution must bring about innovation in the following areas: the content of technological processes; industry structure and manufacturing locations; the internal structure of production and avenues for its cooperation and integration with science and education; and economic relations and institutions that ensure the progress of a radically new type of material production. Reforms should encompass all elements of the production process, including its organization, technology base, product, and, of course, the nature and quality of industrial labor.

The sixth technological mode is characterized not only by more in-depth knowledge of a product, but also by the interaction of various types of knowledge and, accordingly, the types of technologies applied in the production of the product. Paramount importance is assigned to the integration, convergence, and interdependence of nanotechnology, biotechnology, information technology, and cognitive science (NBIC). This phenomenon is commonly referred to as the NBIC convergence.

Chapter 2: Trends in Socioeconomic Developmental Goals and Priorities emphasizes that, on their own, trends in technological shifts are not as important for the strategic vision as results in the field of production caused by such shifts. This results in two pivotal trends: the expansion of opportunities to satisfy human wants (including opportunities made available by the synergy of technologies) and a decreased reliance on humans from immediate material production. To transition to new stages in the technological progress, it is necessary to increasingly master new knowedge and find ways for its technological application. It is precisely knowledge-intensive technologies that are becoming the most advanced.

The growing significance of information technologies does not show a decline in the defining role of material production, but instead leads to a continuous increase in the knowledge intensity of material production. Knowledge-intensive production allows for the faster satisfaction of growing wants. An increase in the level of new technologies allows for a decrease in capital, resources, and energy consumed in the process of production, which potentially creates the opportunity for lowering the share of resources at the expense of satisfying a standard unit of human wants.

Thus, there comes a point when the knowledge component becomes much larger than the material part in many goods. We are already witnessing the emergence of Industry 4.0 and smart factories that work closely with the Internet of Things, or, rather, the Industrial Internet of Things, which allows both for the interaction of autonomous technical devices between each other and for human control over them. Here, we see the precursor of a different machine-based, industrial—but at this point "unmanned"—production.

The new industrial state (described by J.K. Galbraith) is becoming a thing of the past; it is a failed journey after the post-industrial future. We are observing the genesis of the new industrial society of the second generation (NIS.2), which must resolve the preceding era's contradictions. Production should be driven by human intellect as opposed to the desire to boost consumption, gain prestige, or accumulate capital. But to achieve that the human mind itself must evolve so as to change the currently established hierarchy of values.

As society transitions to the new industrial society of the second generation, changes in the technological foundation of production will inevitably cause changes in the system of economic relations and institutions. A new industry determines the need for reshaping the market and state regulation, as well as private entrepreneurship and public property formats. The base value of product acquisition will sharply decrease under the NIS.2 driven by completely new and virtually unlimited opportunities for the satisfaction of real human wants. We can predict with a higher degree of surety that the economy of joint ownership and and diluted ownership rights will dominate the NIS.2 stage. At the same time, technological forces will require the human intellect's direct and total control, which, in turn, will progress toward setting fruitful—as opposed to destructive—development trends.

In the absence of a rational separation between real and feigned wants, technological forces are capable of not only critically changing certain human characteristics with both natural opportunities for and restrictions on consumption, but also of corrupting human nature itself. This trend promotes a new consumer type engulfed in an eternal unscrupulous quest for fictitious goods. This then causes increasing pressure on the Earth's resources despite all of the opportunities for lowering the resource intensity of production. Such risks have already materialized, and any development strategy must suggest ways to avoid growing threats to the human civilization.

If we do not supplement our growing knowledge in the field of technology with other knowledge on the relevance of reasonable self-restriction, the application of noo-approaches to organizing our lives and, particularly, opportunities offered by technological progress, we will clearly arrive at a catastrophe. The noo-approach stipulates the adunation of technological might with the power of human knowledge encapsulated in cultural traditions.

Production will overstep the boundaries of technology and enter the realm of human intellect reliance, however, on strictly material processes of noo-industrial production, for the intellect could neither exist nor develop in the absence of a link with such processes!

Noonomy serves as a basic element of the noo-society, a global "nomos" (principle, structure, order, etc.) that determines the noneconomic method for organizing human economic activity and satisfying human wants with an emphasis on cultural imperatives, as opposed to economic rationality.

Noonomy does not constitute a strategic goal per se. The goal is the pursuit of development priorities embedded in the noonomy, particularly personal development and the formation of homo culturalis—both as the main objective of production and the main factor of its progress.

The strategy of progress toward the noonomy should consider that this path comprises more than one stage. First it reaches an interim stage characterized by not quite the disappearance but the shrinkage of economic forms and institutions that mediate the satisfaction of human wants. After that, humans are ultimately withdrawn from immediate material production, cease their immediate labor activities in this segment, and concurrently abandon economic rationality and transition to the noonomy.

Capital always pursues the expansion of mass production and mass sales and distribution, which engenders constant production development, improvement of technologies, progress of productive forces and simultaneously an increase in the variety of human wants. But since the nature of these wants makes no difference from the perspective of economic rationality, we are seeing the development of false wants that prey on human weaknesses. The modern market economy resorts to bolstering simulative wants in its pursuit of sales volume, which constitutes an essential component of the growing resource burden on the environment.

But can society set the restriction of human wants as one of its strategic goals? Or perhaps, on the contrary, is the goal to ensure their full satisfaction?

These two approaches do not pose an alternative in the case when (1) there is an internal as opposed to external restriction of wants, i.e., when humans exercise the self-restriction of simulative wants; and (2) insofar as self-restriction pertains specifically to simulative wants, opportunities for the satisfaction of reasonable wants increase.

That becomes possible during the transition to nooproduction, which, to a great extent, promotes human's rather than manufacturer's material conditions for their existence. The need for self-development, spirituality, communication, and public recognition will gain paramount importance.

As humans are withdrawn from immediate production processes and their functions focus on control and goalsetting, human activity shifts toward predominantly creative functions related to the discovery and technological application of new knowledge. Such humans prioritize wants that are related to personal development and constitute a prerequisite for the development of creative potential.

Economic rationality is becoming increasingly dubious due to its fallout. It deforms the structure of human wants by trying to make them fit within the tight boundaries of monetary symbols of success and by assigning the status of rationality only and precisely to those achievements that result in higher value metrics that are shaped by the market. But the change in the content of human activity and the nature of wants satisfaction is accompanied by a change in the criteria for the rationality of consumption and, therefore, the structure of wants. Economic rationality criteria are being replaced by criteria that assess the reasonableness of wants as determined by human culture.

Thus, production no longer seeks to ensure quantitative augmentation of the volume of consumption. Its goal shifts to ensuring the quality of life. Transition to noo-production is related to a dramatic change in understanding the quality of life as a concept targeting human development and establishing the prominence of the human personality. This goal does not reject the diversity and variety of human wants, nor does it stipulate the renunciation of consumption and adoption of an ascetic ideology. On the contrary, it can be achieved only by developing human universality both in the field of production activity and consumption. The only shift pertains to criteria for assessing human wealth. These criteria depart from economic rationality and adopt rationality that is determined by culture.

Chapter 3: The Basis for Strategizing National Development discloses problems that arise during the transition from general theoretical foundations of the strategic vision to the formation and implementation of a specific strategy. Even when such conjectures rely on a theory that ensures a long-term perspective, it would be difficult to substantiate new strategic perspectives, select priorities, and develop scenarios when the past only partially extrapolates into the future and future social processes and economic agents remain largely unknown.

The theory of noonomy does not constitute an instrument for calculating the exact timeline for the onset of turning points in the development and achievement of respective goals. The theory of noonomy, however, allows for determining the logical interrelation of events and, consequently, the progression of movement toward these goals.

The vision of goals stems from a certain understanding of our interests and values that ensues from the theory of noonomy. First and foremost, these involve values of personal development reliant on the entire body of riches offered by human culture. The theory of noonomy considers these factors as pivotal for the progress of public development.

Continuous tracking of trends in science and technology acquires paramount importance for implementing a strategy based on the theory of noonomy. Assessing the effect of these trends on the public order, natural environment, and humans themselves is also extremely significant. The insight into the future that is developed under the theory of noonomy allows for discerning long-term development trends and linking them to specific steps in science, technology, economy, management, culture, and many other areas.

Movement toward the noonomy is not a single strategic project, but a strategic vision that encompasses a succession of such projects, whereas the most distant ones still cannot be perceived in detail. A strategy should extend much further and deeper than what is apparent to everyone. A strategy seeks to ensure the efficient movement of the object of strategizing to the reality that does not exist yet and will only start to take shape by a certain period determined by the strategy's horizon.

That is the approach suggested by the theory of noonomy. From the perspective of this theory, Russia's development strategy should be based on a radical abandonment of the current state. Strategic noonomy-based thinking about paths of societal development, including Russian society, should be ahead of common notions by at least several decades and go beyond the life span of one generation.

The theory of noonomy by itself does not suffice for the development of a holistic Russia's national development strategy for the foreseeable future. It is necessary to use general theoretical perceptions about principles of strategizing and paths in the evolution of human civilization as the foundation for developing an entire complex of strategic solutions with various time horizons and degrees of specificity. The mission, vision (including principles and priorities ensured by competitive advantages), and the goals per se (placed on a timeline) jointly form the concept of a strategy. Strategizing incorporates the process of development, long-term implementation, monitoring, and subsequent refinement and updating of a strategy. Moreover, when developing a strategy, it is necessary to ensure and utilize the interrelation between foresight, forecasting, strategizing, and long-term planning.

Chapter 4: Strategic Goals of Socioeconomic Development shows how development goals—which lay the foundation for strategic projects that facilitate the realization of these goals—are formulated based on a national development mission and relevant vision of the future. The definition of a mission reflects national values, interests, and priorities, and provides the platform to consolidate the society's achievement of strategic goals. Russia's national development mission ensues not only from the desires of people living in our country, but also from objectively conditioned trends that underlie the need for increasing the quality of life and creating conditions for developing human potential.

Strategy goals and the identification of national development landmarks directly depend on the answer to the following question about our mission: What do we want our country to be like, and what position do we want to assume in the global system?

Seeking to become a global leader, Russia should set the goal of achieving a new qualitative state of society as predicted under the theory of noonomy—i.e., transition to the NIS.2. An interim goal that is a prerequisite for nearing the NIS.2 should stipulate Russia's reindustrialization on the latest technological foundation.

A new wave of technological changes lies ahead, as seen in the unfolding industrial and technological revolution, and new leaders will be forged from those who can ride this tenth wave. It is essential that we understand that growth—whether in the GDP, profit, or consumption, etc.—is neither the only nor the most important thing we need. Development is much more important.

The implementation of each strategy goal warrants the development of a target program. A target program establishes chronologically interrelated tasks secured with all the necessary resources. The resolution of these tasks allows for achieving the set goal. While a goal serves as a qualitative reference for the realization of a strategic priority, a program constitutes an element of a strategy with substantiated quantitative characteristics.

In Russia, the most important target programs in critical areas of development are currently taking the shape of national projects. It is necessary to convert national projects into subordinate elements of the national strategy that ensue from the definition of the national mission, strategy goals, national priorities, and the scope of tasks required for promoting these priorities.

To achieve that, the Russian economy requires rather substantial systemic changes, including the implementation of an active industrial policy and the transition toward economy management based on long-term strategy, medium-term indicative plans, and programs reliant on scholarly foresight. The government should guarantee state paternalism for long-term investment into R&D and technological upgrades, provide consistent tax support, and ensure affordable and convenient loans targeting the real sector of the economy (and particularly high-tech production).

Chapter 5: Priorities for the Modernization of National Economy substantiates the need for accelerated technological modernization of the Russian economy as a platform for reindustrialization that provides material premises for boosting the quality of life.

Technological progress brought on a gap between the structure of the economy, which had already become obsolete, and new technological opportunities. The stage of transitioning to a new development paradigm has begun; the transition has already started and old models and methods for overcoming crisis phenomena have lost their relevance. The current model of the global capitalist economy is slowing down qualitative revolutionary shifts in the technological foundation, thus creating a window of opportunity for Russia.

The main goal of reindustrialization as an economic policy stipulates that Russia restore the role and place of industry as a core component of the national economy during its restructuring and ensure prioritized development of material production and the real sector of the economy based on a new advanced technological mode under national modernization.

Rather than developing competition in general, economic modernization requires creating conditions where Russian entrepreneurs will have to use technological modernization as the main instrument of competition. It is necessary to convert the process of creating new technologies into an uninterrupted flow. That is not possible unless we restore efficient interaction between production, science, and education reliant both on past Soviet and current international experiences.

Russia's National Technology Initiative (NTI) as an instrument of modernization targets the development and application of technologies for which Russia has scientific and economic potential. But it is equally important to determine export potential for promising high-tech production. It is essential that innovations emerge not of their own accord, but in response to existing demand and that they help satisfy customer needs.

The Russian industry is dominated by the fourth and fifth technological modes, so our current level of technological development requires at least an active industrial policy and strategic planning under the market economy. A strategy cannot be implemented in the absence of specific plans and programs. A strategic plan is a critical element and main administrative instrument of strategy implementation. It is impossible to imagine a future society, an intellectual society, a noo-society, without the institution of planning as a core, basic instrument of public administration, as well as its entire existence.

Our national experience demonstrates that eliminating strategic planning from main instruments of public administration yields development that is dominated by inertia and the impossibility of overcoming path dependence. This in turn hampers the correction of profound structural imbalances, leads to accumulation of systemic risks, and negates the possibility of achieving any ambitious goals.

Strategic planning relies on constructing a vision of the future that is not just based on the correlation of wants and resources required for their achievement; first and foremost, it stems from strategic goals and public priorities. To determine such goals and priorities, the process of strategic planning relies on broad scientific expertise and on engaging representatives of public opinion and business interests. They are expected to participate both in strategy development and in subsequent plans, projects, and programs.

Assessing the situation in aggregate, we must admit that Russia still lacks a general national strategy. We have yet to create such a strategy and ensuing strategic plans.

Chapter 6: Strategizing on the National, Regional, and Sectoral Economies uses specific examples to show the formation of such components of a national development strategy as regional and sectoral development strategies.

Strategizing for regional development requires the resolution of a rather controversial issue. The strategic goals of regional development should correlate with goals that are set at the national level. At the same time, the difference in regional circumstances requires dramatically different approaches to determining regional development strategies.

Regional strategies that are currently adopted in Russia, as well as ensuing regional development plans and programs, have been typically constructed not as a system, but as a simple list of measures targeting the resolution of certain issues that seemed important at the time. The failure to account for territorial and spatial aspects of regional development strategies reflects on the increasing imbalance in territories' development. An analysis of experiences with the development of even the most successful documents on regional strategic planning in St. Petersburg and Primorsky Territory indicates that there is a wide range of issues that have not been fully resolved.

Strategizing at the regional level is determined both by a general vision of strategic perspectives and development goals as defined in the context of pursuing economic reindustrialization and a vision of specific sectors' contribution to resolving this task.

When providing terms of reference for the engineers of sectoral (and regional) development strategies, various government bodies have so far failed to ensure a true integration of such strategic plans and programs, starting with the level of methodology for their development. Such a situation inevitably follows from Russia's lack of a mature development strategy. This leads to a deficit of clearly formulated quantitative and qualitative criteria, which government agencies need to perform continuous monitoring of a strategy's implementation.

An analysis of sectoral development strategies for the food, tourist, and machine tool industries shows that they have strategic prospects and exposes the aforementioned shortcomings in the formation of strategic projects.

In the Conclusion, the authors emphasize that they sought to not only present conceptual terms of the theory of strategy and noonomy, but also to invite economists, sociologists, culturologists, and even philosophers to heed a new aspect of a promising interrelated study of humans, their creative activity, and the natural environment.

Contents

A Word To The Reader .... *xxi*

1. **Global Development Trends** 1
    1.1 Progress in Science and Technology and a New Technological Order
    1.2 Industrial and Technological Revolution
    1.3 Transformational Processes in the Global Economy and Civilizational Development: Opportunities and Risks
    1.4 Noonomy as a Conceptual Platform for the Global Transformation of Society

2. **Trends in Socioeconomic Developmental Goals and Priorities** 61
    2.1 Change in the Nature of Needs in the Movement Toward Noo-Production
    2.2 Quality of Life as a Target for Societal Development

3. **The Basis for Strategizing National Development** 79
    3.1 Strategizing as a Method for Identifying Development Interests, Priorities, and Goals
    3.2 Strategic Goal-Setting and Planning Tools

4. **Strategic Goals of Socioeconomic Development** 107
    4.1 Identifying Strategic Targets
    4.2 National Projects

5. **Priorities for the Modernization of the National Economy** 115
    5.1 The Acceleration of Scientific and Technological Growth for the Purpose of Achieving a New Quality of Life
    5.2 Reindustrialization, Digitization, and the National Technological Initiative

6. **Strategizing on the National, Regional, and Sectoral Economies 147**

    6.1 Development of Opportunities for Strategizing the Transition of the National Economy to the NIS.2

    6.2 Practical Experiences in Strategizing Regional and Sectoral Development

Conclusion *177*
Bibliography *179*
Appendix: New Ways Ahead for National Economy
    (Regarding Possible Russia's Development Strategy) *193*

A Word To The Reader

This book is dualistic in its nature. It seeks to combine two approaches to the analysis and assessment of societal development prospects and to strengthen the capacity of each.

One approach has at its core managing the information and technological development of society—its social and economic transformation—through developing and implementing a particular strategy with a concept or doctrine of the planned guidelines as its first stage. Strategizing the information-technological transformation of society is proved to be most effective when it covers long-term development periods. The accumulated strategizing practice is mostly related to 10–15-year periods. However, strategy development and implementation examples for 50, 100, and even 200 years also exist. Such long-term strategies lead to significant and even fundamental changes in the values and priorities of socioeconomic development.

Philosophers, sociologists, economists, technologists, and members of many other professional fields, therefore, benefit from developments in the long-term qualitative change of society. Another approach described in this monograph, which is implemented in conjunction with strategizing, is connected to the conceptual understanding of long-term development. The concept of noonomy represents a complex theory of transformation based on technological change and the resulting shifts in social organization. It demonstrates not only trends but also qualitative social shifts to which these transformations lead. In this way, the approach put forward in the theory of noonomy makes it possible to anticipate and evaluate distant horizons of social development and to grasp the transitions from one stage to the next. Employing the concept of noonomy in the processes of strategy is a prognostic phase, immediately preceding the processes of strategy and creating a reference point for them.

But a vision of the future is not valuable in and of itself. It becomes socially relevant when it becomes a tool and a technology that allows the use of this knowledge, adjusting the evolution of society toward what is objectively possible and appropriate. This approach avoids moving toward unreasonable, unattainable, or socially dangerous benchmarks. In turn, strategy techniques turn knowledge about the future into a tool for the effective management of development processes.

The concept of noonomy is particularly suited to a strategic approach, as it involves anticipating events that are not obvious to "common sense" and which look far beyond the current inertial agenda. The direction of this inertia, based on the extrapolation of previous scenarios, is possible without special strategic efforts. A sound strategy allows taking an unobvious path into an uncharted future.

A theory that allows us to see the emerging changes and the future that follows the qualitative shifts in global development should therefore be used to form a strategic vision of society's destiny and the strategic impact on its development.

This book represents the first time strategy concepts (V. L. Kvint) and noonomy (S.D. Bodrunov) have been brought together. The idea of uniting the authors' views on the problems of civilizational development has a common scientific platform: the definition of long-term goals and the choice of economic and strategic tools to achieve them.

This book summarizes the authors' main approaches to the issues at hand. This facilitates solving the applied problem set by the authors, which is to demonstrate the productivity of synthesizing these approaches to the study of societal development patterns for subsequent use in their theoretical and practical implementation The presentation chosen by the authors promotes an understanding of the concepts in their entirety.

CHAPTER 1

Global Development Trends

A strategic approach that looks far beyond the horizon of current events and trends is not based solely on the strategist's intuition or some visionary gift. Any strategy, no matter how long-term, is based on real-life facts, on the objective processes that determine the development of society and its various sub-systems. In contrast to the usual view, the strategic view captures in the current course of events the regularities that define the coming changes. And it is these profound shifts in societal development that shape the strategy's goals.

The theory of noonomy concentrates on the study of those processes and trends that naturally lead us to a new stage of society. It does not extrapolate from these trends but studies them as a starting point for understanding how these trends will lead to the inevitable qualitative shifts that shape future reality.

The development of material production is a determining factor in social development, with its image being shaped by progress in knowledge and its technological application. As Nobel laureate Edmund Phelps remarked, "modern economics, if you understand it as a vast and ongoing project to invent, develop and test new things and methods that people can work and enjoy, has had a profound effect on work and society."[1] Therefore, studying the processes and trends in this particular area should commence with understanding both the goals of strategizing and the means needed to achieve these goals.

The exhaustion of the previous stage of technological development and the accumulation of prerequisites for a transition to the next stage of technological development are distinctive features of the current state of the world economy. Predictions of the upcoming technological and industrial revolution have been raised so often recently for a reason.

---

[1] Phelps, E. *Mass Prosperity: How Grassroots Innovation Became a Source of Jobs, Opportunity and Change.* Gaidar Institute Publishing House; Liberal Mission Foundation, 2015, 62.

What shifts in technological development will this revolution be based on and what effects will it produce?

1.1 PROGRESS IN SCIENCE AND TECHNOLOGY AND A NEW TECHNOLOGICAL ORDER

1.1.1 THE SIXTH TECHNOLOGICAL MODE

Modern industrial production is heading toward the sixth technological mode. Biotechnology, nanotechnology, and artificial intelligence are at the forefront of technological progress; digitalization is spreading to cover many areas of human life. These technologies have the potential to revolutionize the way we live and the quality of our lives. We can already see the first shaky contours of a future technological reality. Experts suggest that if current development tendencies continue (and it should be noted that crisis processes make adjustments), the sixth technological mode will become the basis for deploying *a new scientific and technical, technological, and industrial revolution* in 2020–2025. This revolution will be founded on the synthesis (convergence) of many cutting-edge technologies, some of which are still in their infancy.

> *The sixth technological mode* is a complex of technologies that is currently being established, including nano-, bio-, information and cognitive technologies, distinguished by the convergence of technologies and the formation of hybrid technologies with the integrating role of information technologies (digitalization, artificial intelligence, and big data processing).

Everyone should keep in mind that changes in material production will be *systemic* and *holistically interrelated* when determining the strategy for industrial development. Let us emphasize some of the major points to be considered when *creating a new industrial system at the cutting edge of twenty-first century science and technology*.

This is how we envisage the industry's main features:

- updating the content of technological processes;
- changing the structure of industrial enterprises (micro-level);
- changing the sectoral structure of industry (macro-level);

- changing approaches to organization/localization of production facilities;
- forming new types of industrial cooperation;
- increasing integration of production with science and education;
- transitioning to the ideology of "continuity" within the innovation process in production;
- forming economic relations and institutions aimed at industrial/scientific-technological progress.

The following should become novel: *the content of technological processes*; the *structure* of industries and *distribution* of productions; the internal *structure* and *types of production cooperation* and their integration with science and education; and economic *relations* and *institutions* ensuring the progress of fundamentally new material production.

It is not sufficient to master the technology to manufacture products that meet today's requirements. These new standards should be extended to the areas of quality management, production management, logistics, and human resources. The changes concern *all elements of the production process*: its *organization*, the *technological base*, the *product,* and of course, the *nature* and *quality* of industrial work. In the changing *nature* and *forms of organization* of industrial production, for example, we should look at the trend toward the *individualization of production*, which has been developing since the end of the twentieth century, and toward the organization of work for the *individual consumer.*[2]

The major technological challenges of the twenty-first century industry include:

- "increasing development of new technologies that improve productivity and make production cheaper;
- increasing "individualization" of production, applied technologies and manufactured products;
- introducing the principle of product modularity;
- accelerating intellectualization, computerization, and robotization of production;
- developing network technologies and implementing the network principle of production organization;
- miniaturizing and compacting production;
- strengthening the trend toward low-cost and zero-waste production;

---

[2] Bodrunov S. D.; *Noonomy,* Cultural Revolution, 2018, 75–76.

- permanently increasing the rate of technology transfer;
- increasing tendencies of "physical" rapprochement of developer and manufacturer, and a reduction in time for the implementation of new products;
- expansion of "intellectualization zones" of labor;
- "clustering" of production relations;
- increasing role of individual, motivational, psychological, social, and other characteristics of participants in production activity;
- reducing the share of labor costs in the industry for the production of new products with an increase in the cost of their development;
- changing the structure of production profitability in favor of knowledge-intensive and highly processed products."[3]

The most significant of these challenges is "the principle of the individualization of production with simultaneous *modularities* for high-tech sectors, such as machine tools, aviation (civil and military), heavy machinery, etc.

The individualization of production and the establishment of a contact between the producer and the individual consumer are underpinned using modern information and telecommunication technology.

The development of the Internet has led to the mass formation of B2B and B2C communication platforms. This established an effective tool for direct interaction between customer (consumer) and manufacturer. Soon this, combined with the development of groundbreaking new technologies (virtual design, computer visualization, 3D printing, etc.), will make it feasible to create *individual* industrial products, virtually *waste-free*, and deliver them to the consumer almost *instantly*."[4]

At the same time, the "individualization of production facilitates *the transition to a network-based organization, not only of business but also of material production*. This permits the quick creation and reconfiguration of interactions between producers and sub-suppliers, and more generally with subcontractors and outsourcers. On this basis, it is possible to quickly adapt the product to the individual demands of consumers and then move on to new products aimed at other consumers or users, other markets,

---

[3] Bodrunov, S. D.; What Kind of Industrialization Does Russia Need? *Economic Revival of Russia* 2015, *No.2* (44), 11.

[4] *Ibid.*, 12.

etc. Network organization promotes more and more individualization of production, and these processes take on an *avalanche-like character.*"[5]

1.1.2　NBICS TECHNOLOGY CONVERGENCE AND HYBRID TECHNOLOGIES

The sixth technological mode is characterized not only by the increased knowledge capacity of a product but also by the interaction of various types of knowledge and, accordingly, the technologies used in producing any product. The integration, convergence, and mutual influence of information, bio- and nanotechnologies, and cognitive science appear most prominent. This phenomenon is called *NBIC-convergence* (i.e., nano, bio, information, cognitive). The term was introduced in 2002 by Michael Roco and William Bainbridge, authors of the most significant work in this area to date, *Converging Technologies for Improving Human Performance*[6] prepared by the World Technology Assessment Center (WTEC).[7]

> *NBIC-convergence*—the mutual penetration of nano-, bio-, information and cognitive technologies, leading to the creation of technological processes in which these technologies function as mutually supportive and form an inseparable whole.

According to J. Spohrer, "The same report proposed the notion of NBICS convergence, which includes the social sciences.[8] Although this

---

[5] *Ibid.*

[6] As defined in this report, NBIC convergence is the "synergistic combination of four major NBIC (nano-bio-info-cogno) provinces of science and technology." *See:* Overview: Converging Technologies for Improving Human Performance; *Converging Technologies for Improving Human Performance: Nanotechnology, Biotechnology, Information Technology and Cognitive Science,* Roco, M.; Bainbridge, W.; eds., 2004, 1.

[7] Pride, V.; Medvedev, D. A.; The Phenomenon Of NBIC-Convergence: Reality And Expectations. *Philosophical Sciences* 2008, No.1, 97–98.

[8] Spohrer, J.; NBICS (Nano-Bio-Info-Cogno-Socio) Convergence to Improve Human Performance: Opportunities and Challenges. *Converging Technologies for Improving Human Performance: Nanotechnology, Biotechnology, Information Technology and Cognitive Science,* Roco, M.; Bainbridge, W., eds., WTEC, 2004, 102. http://www.wtec.org/ConvergingTechnologies/Report/NBIC_report.pdf.

approach is widespread in Western and domestic scientific literature,[9] social sciences do not make a significant contribution to solving the problems of convergent technologies' development and application yet."[10] Social technologies are only really used to develop artificial intelligence systems designed to interact with the consumer (or rather manipulate the consumer). Humanists write about the social problems posed by new technologies more often than they do about integrating social knowledge in their development.

As S. Borunov states, "Considering the interconnectedness of sixth mode technologies and the interdisciplinary nature of modern science, we are ready to discuss the expected (in the long term) merging of the NBIC fields into a single scientific and technological field of knowledge. Almost all levels of matter organization will be the subject of study and action: from the molecular nature of matter (nano) to the nature of life (bio), mind (cogno), and information exchange processes (info)."[11]

"So, the distinguishing features of NBIC convergence are:

- intensive interaction between the mentioned scientific and technological fields;
- significant synergistic effect;
- breadth of the subject areas considered and affected, from the atomic level of matter to intelligent systems;
- identification of the prospects for qualitative growth in technological capabilities for individual and societal development."[12]

The convergence of technologies within the sixth technological mode has resulted in the extensive use of hybrid technologies, where different combinations of machine and non-machine technologies, together with information technology, are used as tools to manage and direct natural processes to achieve mankind's desired goals, opening the door to a new technological revolution.

---

[9] Kovalchuk, M. V.; Convergence of Science and Technology–Breakthrough to the Future. *Russian Nanotechnologies* 2011, V. 6, 1–2, 21. http://www.nrcki.ru/files/pdf/1461850844.pdf; Kovalchuk, M. V., Naraikin, O. S., Yatsishina, E. B.; Convergence of Science and Technology and the Formation of a New Noo-Sphere, *Russian Nanotechnologies* 2011, V. 6, 10–13.

[10] Bodrunov, S. D; .Convergence of Technologies—a New Basis for Integration of Production, Science and Education. *Economic Science of Modern Russia* 2018, No. 1, 12.

[11] *Ibid.*

[12] Pride, V.; Medvedev, D.A.; The Phenomenon Of NBIC-Convergence: Reality And Expectations. *Philosophical Sciences* 2008, *No. 1,* 104.

Hybrid technology—the combination of two or more technologies of different types in one device to achieve one useful result. Often this combination is *convergent*, with one technology supporting the other to some extent.

On March 31, 2019, the Google search engine returned 17 million links for the query "гибридные технологии" (Russian for "hybrid technologies"), and 641 million for the same search query typed in English. They mention hybrid technology in the field of industrial processing, the automotive industry, artificial intelligence, seed pre-sowing, electronic security, nuclear desalination, the military, machine translation, cardiac surgery, etc. It is challenging to conceive of an area where hybrid technology has not been applied. However, no general definition of hybrid technologies was found on the Russian-language Internet. In the English-speaking segment, there is such a definition on one of the websites dedicated to climate technology: "Hybrid technology systems combine two or more technologies with the aim to achieve efficient systems."

1.1.3 ADDITIVE AND DISTRACTIVE TECHNOLOGIES

The sixth technological mode does not go beyond the industrialized mode of production, even with a significant increase in the role and importance of non-machine technology (bioengineering, etc.). Convergent (hybrid) technologies, on the other hand, give a second life to the industrialized mode of production, combining machine and non-machine principles of impact on nature to create products that satisfy human needs, with the least amount of materials.

3D printing technology, based on new types of machine technology (printers) integrated with information technology and virtualization (3D modeling), offers great opportunities. This will probably lead to a dramatic expansion of additive technologies and a reduction in the weight of the traditional manufacturing industry. Processes of "assembling" products from elements by combining or building up material (usually layer by layer) to create an object based on a 3D model replace "processing" of source material employing distractive ("cutting") production techniques (trimming, chipping, sawing off material from the workpiece).

Among the traditional industrial technologies are those classified as additive: casting, sintering building materials, and powder metallurgy. Today, the capabilities of these technologies are being combined with 3D printing capabilities. We are witnessing the creation of 3D printers capable of printing entire buildings and structures, or at least large blocks of them. Building houses from 3D-printed parts is already a reality, for example, the first house has been printed on a Russian-made printer in Yaroslavl.[13] An entire office hotel in Denmark was printed on the same Spetsavia printer.[14]

Additive technologies encompass a range of processing techniques (extrusion and jetting, sheet lamination, photopolymerization, powder synthesis, and direct point energy release) and use a wide variety of materials (plastics, new plastics, metals, composites, flexible materials, materials for metal casting processes, ceramics, etc.).[15]

Today, 3D printing technologies are already being integrated with biotechnology to generate human organs for transplantation on 3D printers. So far, only bioprostheses (implants) made of artificial materials to replace bone and cartilage tissue, as well as prostheses of the hand, are actually used. However, experiments on growing tissues of human organs (liver, kidney, bladder, skin) are already practically used for testing pharmaceuticals.[16] There is no doubt that the future lies in these technologies.

1.1.4 THE ROLE OF INFORMATION AND COGNITIVE TECHNOLOGIES

In the development process, people move toward an awareness of their escalating needs and ways of addressing them. Knowledge, by its very nature unlimited, reveals to man not only the answer but also a wider horizon, creating new needs. This horizon is limited at each stage of cognition only by a person's current ability to grasp it. This reflects all human development, including scientific and technological progress and the development of social relations.

Thus at one stage mechanic forces were recognized, understood, and put into production, followed by the much more knowledge-intensive

---

[13] Hybrid Technology. UN: Climate Technology Centre and Network (CTCN). https://www.ctc-n.org/technologies/hybrid-technology (accessed May 27, 2022).

[14] Europe's First Residential Building Printed on a 3D Printer Was Presented in Yaroslavl. https://specavia.pro (accessed June 8, 2022).

[15] The Construction of Europe's First 3d Printed Building Has Begun. https://3dprinthuset.dk/europes-first-3d-printed-building (accessed June 8, 2022).

[16] For an overview of the possibilities of additive technologies, *see:* Prosvirnov, A.; New Technological Revolution is Passing Us By. *ProAtom Agency,* December 11, 2012. http://www.proatom.ru/modules.php?name=News&file=article&sid=4189 (accessed June 8, 2022).

forces of electricity, and now we have information and cognitive resources as our base.

All of this is only achievable through digitized computer control integrated into the technological processes themselves, which entails the widest possible use of information and communication networks. And this is different from the "digitalization" imposed on traditional technological processes in the fifth or fourth technological modes. Separate the program control unit from the CNC machine, for example, and you get a conventional machine tool. But try to do the same thing with a 3D printer, and you get a dysfunctional machine. Try disabling Industry 4.0 from the Web and you will stop entire industries.[17] The current trend toward "digitalization" of the economy is also possible beyond the sixth technological mode. However, the sixth technological mode makes it not only economically feasible but also technologically predetermined. Without the use of information technology and information communication in digital form, NBIC convergence is not an option.

> *Digitalization* is a set of solutions associated with using modern information and communication technologies (Internet, mobile communications, big data processing, artificial intelligence, etc.) in a predominantly digital form.

"Cognitive technology in the sixth mode, using self-learning artificial intelligence (AI) systems, is also penetrating areas where there was previously no alternative to the use of human labor. AI systems can do this by searching, storing, sorting, and collating information so that decisions can be made on this basis.

It is cognitive technology, using biotechnology and information and communication technology, which allows direct human interaction with unmanned technological processes (human–machine interfaces, human–machine systems, and human–machine networks).[18] On this basis, robotics production is being revitalized, becoming more flexible, more adaptable, and more productive. AI is still quite far from being capable of discovering

---

[17] Bioprinting of Organs on a 3D Printer, How Does It Work? Make3D, https://make3d.ru/articles/biopechat-organov-na-3d-printere (accessed June 8, 2022). *See also* posts on the 3d Bioprinting Solutions portal: Interview by Yousef Hesuani (November 8, 2017); Company Staff reports at the Annual Biofabrication Conference in Beijing (October 27, 2017).

[18] Bodrunov, S. D.; Convergence of Technologies—A New Basis for Integration of Production, Science and Education. *Economic Science of Modern Russia* 2018, No. 1 (80), 14–15.

new knowledge (it can *receive* it by storing and analyzing existing information and it can *transmit* it via ICT, but it *cannot* discover knowledge). This is the reason why the new technological mode imposes new and increasing demands on the research and cognitive activity of the human being. Thus, technology convergence approaches require interdisciplinarity in the organization of scientific research. Convergence in education must correspond to the orientation toward convergent technologies. This is still largely hampered by the sectoral organization of both science and education.[19]

But why do these trends in technological progress constitute a new technological mode? And what defines the transition from the previously observed co-existence and interaction of different technologies to their convergence, i.e., the formation of hybrid technologies?

As S. D. Bodrunov writes, "To answer these questions, attention should be paid first and foremost to modern information technology and the related process of "digitalization" of other technologies. Information and communication technologies, unlike any other, demonstrate the ability to penetrate any technological process, and digitalization becomes the technological platform capable of combining heterogeneous technologies into hybrid technological processes."[20] M.V. Kovalchuk specifies, "Information technology has become a kind of 'hoop' that unites all sciences and technologies."[21] Therefore, digital technology is at the core of the new technological mode.

The other technologies that make up this mode "have in common, on the one hand, their ability to converse with each other, and on the other hand, the fact that this convergence aims to fulfil two major trends characteristic of the current stage of technological development. This is, firstly, the development of a trend toward the displacement of human beings from the indirect process of material production and, secondly, a trend toward a

---

[19] For a review on this topic, *see:* Milena Tsvetkova, Taha Yasseri, Eric T. Meyer, J. Brian Pickering, Vegard Engen, Paul Walland, Marika Luders, Asbjørn Følstad, George Bravos. Understanding Human-Machine Networks: A Cross-Disciplinary Survey [Online], Cornell University Library. https://arxiv.org/pdf/1511.05324v1.pdf

[20] Bodrunov, S. D.; Convergence of Technologies—A New Basis for Integration of Production, Science and Education. *Economic Science of Modern Russia* 2018, No. 1 (80), 15–16.

[21] *Ibid.,* 13.

sharp increase in the knowledge content of the product and a corresponding decrease in the proportion of material inputs in its production."[22]

Other researchers confirm the importance of this trend: "Technology is basically developing along two vectors: waste-free (the ultimate goal of resource conservation) and humanlessness. This leads to challenges in cleaner production and the development of new control systems—automated, intelligent, flexible production systems based on artificial intelligence, biomechanics, robotics, etc."[23]

1.1.5 TRENDS IN TECHNOLOGICAL PROGRESS AND STRATEGIC DEVELOPMENT BENCHMARKS

The main directions of technological change are thus seen in the growth of knowledge-intensive technologies, the use of hybrid technologies and the increasing role of information and cognitive technologies as a factor that links the development of all other technologies. But what matters more for the strategic vision is not the trends of technological shifts per se but the production outcomes to which these shifts lead. There are two major outcomes: the expansion of opportunities to meet human needs (including through technology synergies) and the progressive technological displacement of humans from direct material production.[24]

A strategy for technological development should not be based on randomly identifying a set of technologies that appear to be "cutting-edge" or "modern." The goals associated with technological change should focus on those trends that fundamentally change the nature of the production process and thus lead to shifts in the social order.

Therefore, to determine strategic interests, priorities, and goals, it is not enough to have a common understanding that new technologies are changing the face of modern production, and even in what approximate direction. It is vital to understand specifically how the characteristics of production are being transformed by new technological trends.

---

[22] Kovalchuk, M. V.; Convergence of Science and Technology—Breakthrough to the Future. *Russian Nanotechnologies* 2011, V. 6, No. 1–2, 14.

[23] Bodrunov, S. D.; Convergence of Technologies—A New Basis for Integration of Production, Science and Education. *Economic Science of Modern Russia* 2018, No. 1 (80), 13.

[24] Sorokin, D. E.; Sukharev, O. S.; Structural and Investment Tasks of Russia's Economic Development. *Economics Taxes Law* 2013, No. 3, 13.

1.2 INDUSTRIAL AND TECHNOLOGICAL REVOLUTION

1.2.1 INCREASING THE KNOWLEDGE INTENSITY OF PRODUCTION

The current stage of global civilizational development is unique. The world is entering not only a new technological mode but also the fourth (sometimes claimed to be the third) industrial-technological revolution. This is in addition to the growing trend toward a new economic mode. In the future, *competitive* economies will be those that can take the lead, not in the extraction and sale of natural resources, but the development and application of *high technology*, and they will ensure the quality of *human capital that can realize it*. The economic leaders of the future will be technological leaders.

It is essential to acquire newer knowledge and find ways to apply it technologically to move to new levels of technological progress. The most knowledge-intensive technologies are becoming the most advanced.

Let us look at current changes in technology, above all those which have become (or are becoming) a reality and are taking place in material production. We should first examine the increasing importance of information technology, rightly identified by "post-industrialist" theorists. However, unlike post-industrialist theory, the theory of noonomy does not see this as evidence of the withering away of the defining role of material production. A different conclusion is drawn from the above-mentioned fact about the continuous growth of the *knowledge-intensity of material production*.

This growth does not record the increased role of information (as many theorists of the information society do),[25] and it is not so much about the *production of information* as about a new type of *material production*.[26] The difference is substantial. As today's global economy shows, the creation of information often translates into the production of information

---

[25] Novikova I. V. *Concept of Employment Strategy in Digital Economy.* Kemerovo State University, 2020.

[26] "Information society" and "knowledge-based society" are a long-standing subject of interest for post-industrialists. *See:* Drucker, P. *The Age of Discontinuity; Guidelines to Our Changing Society,* Harper and Row, 1969; Machlup, F. *Knowledge Production And Dissemination In The United States,* Progress, 1966. (The Production and Distribution of Knowledge in the United States, Princeton, 1962); Masuda, Y. *The Information Society as Postindustrial Society,* World Future Soc., 1983.

noise; economic resources are used to create signs,[27] simulacra,[28] and useful goods rather than contributing to productivity, human quality, and social and environmental issues. Such "informatization" leads to the virtualization of social existence, destroying the human personality, spiritual world, social ties, and the unity of peoples and states.

The *knowledge-intensive* technology of material production is a process that critically synthesizes the achievements of the industrial and information economies. The critical synthesis is manifested, for example, in the fact that in high-tech production, operations and processes play a decisive role in which humans are not appendages to a machine but carriers of knowledge that is then transformed into technology.

In this case, we can talk about the *knowledge intensity* of material production and the *knowledge capacity* of its products.

The main features of this fundamentally *new type of material (knowledge-intensive) production that is taking shape* are as follows:

- a "continuous increase of information and decrease of the material component; miniaturization, [and a] tendency to decrease energy, material, and labor intensity of production;
- such specifics of the *production process* and trends in *technology* as flexibility, modularity, unification, etc.;
- *a network model of structuring*, replacing vertically integrated structures;
- the use of modern methods of production and management (just-in-time, lean production, etc.);[29]
- greening and an orientation toward *new sources of energy*;
- the development of qualitatively new technologies in material production, transport, and logistics (nanotechnologies, 3D-printers, etc.);
- reducing the role of the traditional manufacturing industry (including by replacing manufacturing technologies with additional ones);

---

[27] The issue of knowledge-intensive industry has been debated for a long time. But it also captures the lack of certainty in understanding what a "knowledge-based economy" and "knowledge-intensive industry" are. *See:* Smith, K.; What Is The 'Knowledge Economy'? *Knowledge-intensive Industries And Distributed Knowledge Bases,* Oslo, 2000, 2, 7–9.

[28] Baudrillard, J. *Towards a Critique of the Political Economy of the Sign,* Academic Project, 2007.

[29] Buzgalin, A. V.; Kolganov, A.; The Simulacrum Market: A View Through the Prism of Classical Political Economy. *Philosophy of Economy* 2012, No. 2, 3.

- emphasis on quality and efficiency."[30]

Applying new knowledge in production is an ever-accelerating process due to the growing synergy of beneficial effects (inherent in knowledge as a phenomenon).

As a result, knowledge-intensive production allows for a faster response to growing needs. The new technologies are becoming more advanced, resulting in less capital, materials, and energy consumption, which in the long term will open up the possibility of reducing the unit cost of resources per unit of human need. In this context, exhibitions and fairs play a new role, being transformed into a new economy sector, and becoming a form and method of transferring knowledge-based technologies into production processes.[31]

At some point in many products, the "knowledge" part starts to exceed significantly the "material" one. This conclusion is well illustrated by the graph below, where the curves representing the share of material and intellectual costs in total production costs intersect (Fig. 1.1).[32] Such a moment has already arrived. For example, if you take the iPhone, Apple says that only 4.8% of the cost is attributable to tangible materials. This material/knowledge ratio is common to most high-tech industrial products, clearly marking the arrival of a trend. This will reduce demand for resources, thus changing the position of resource-producing countries in the global economy. From a global resource balance perspective, this is about reducing pressure on natural resources and enabling development while maintaining (and restoring) equilibrium with the natural environment.

A new type of production—knowledge-intensive production—produces a knowledge-intensive industrial product based on knowledge-intensive technology to meet increasing human needs, including, unlike the mass production of generic, first-generation industrial products, the need for innovation. This type of production is not feasible without a

---

[30] For more details *see:* Ohno, T., *Just-In-Time for Today and Tomorrow,* Productivity Press, 1988; Wadell, W.; Bodek, N.; *The Rebirth of American Industry,* PCS Press, 2005; Malakooti, B.; *Operations and Production Systems with Multiple Objectives,* John Wiley & Sons, 2013; Tillema, S.; Steen, M.; Co-existing Concepts of Management Control: The Containment of Tensions Due to the Implementation of Lean Production; *Management Accounting Research* 2015, Vol. 27.

[31] Bodrunov, S. D.; New Industrial Society: Structure and Content of Social Production, Economic Relations, Institutions; *Economic Revival of Russia* 2015, No. 4 (46), 17.

[32] Sadovnichaya, A. V.; *Strategy of Exhibition and Fair Activity,* IPC NRU RANEPA, 2019.

Global Development Trends 15

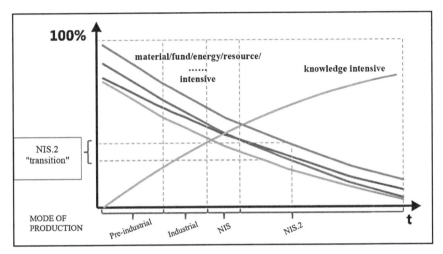

FIGURE 1.1 The process of historical change in the ratio of different product components in production.[33]

high level of knowledge in all its components: materials, labor, organization of the production process and—let us emphasize—the technologies used. Knowledge comes to the fore (and will remain there forever!) in its explicit, "pure" form as the main resource of industrial-technological and social development.[33]

1.2.2 INDUSTRY 4.0 AND SMART FACTORIES: DISPLACEMENT OF HUMANS FROM DIRECT PRODUCTION

The development of the fifth and sixth technological modes has revealed a clear tendency to displace humans from the process of indirect production. The automatic production lines, which were introduced as part of the third and fourth technological modes, were highly specialized and could not be adapted to new production tasks (other technological operations and new products), displacing humans from direct production. They are mainly used in narrow technology niches, for example, in the production of raw materials (chemical industry, rolling mills, paper machines, etc.).

---

[33] This graph was called "Bodrunov's cross" during the discussion at the session of the Department of Social Sciences of RAS (*see:* Grinberg, R. S.; Smart Factories Need Smart People and Smart Economy; *Economic Revival of Russia* 2016, No. 4(50), 155).

Based on the fifth and sixth technological modes, it became possible to produce automated units that could be reconfigured to produce different types of products and technological operations. One of the first visible manifestations of this trend was the production of numerically controlled machine tools (CNC) and universal industrial manipulators (robots). Hardly any non-CNC machine tools are produced in the world anymore. Robot manufacturing, which got off to a strong start in the late 1970s and then slowed down in the 1990s, got a boost from advances in microelectronics and artificial intelligence.

The world's leading countries are paying serious attention to the development of robotization. In the USA and Japan, special groups for robotics development have been established at the governmental level. These are the National Robotics Initiative (NRI) in the USA, founded in 2011, and the Robot Revolution Realization Council in Japan, established in 2015.[34] Japan, the world's longtime leader in industrial robotics and robot production, aims to remain at the forefront. China is challenging this leadership; 154,000 industrial robots were installed there in 2018, which is more than the US and Europe combined. The world now installs more than 400 thousand industrial robots, and in Singapore and South Korea their number is about 7–8% in relation to the number of people employed in the industry.

Modern material production is far removed from the "factory system" (in which humans are an appendage to the machine system) that evolved in the nineteenth century and has survived into the twenty-first century. We are witnessing the emergence of "Industry 4.0,"[35] "smart factories" that work in conjunction with the "internet of things" (more precisely, the industrial internet of things[36]), which provides the interaction of autonomous technical devices and human control over them. The use of embedded sensors and big data processing systems is being developed for this control. Here, we see the prototype of another—machine-based, industrial, but already "unmanned" production.

---

[34] Bodrunov, S. D.; New Industrial Society. Production. Economics. Institutes; *Economic Revival of Russia* 2016, No. 2(48), 11.

[35] Sziebig, G.; Korondi, P.; Effect of Robot Revolution Initiative in Europe – Cooperation possibilities for Japan and Europe. *ScienceDirect*, IFAC-PapersOnLine 2015, 48–19, 160. https://www.sciencedirect.com/science/article/pii/S2405896315026518/pdf?md5=d91 729200da3d63462700e14a0fdefd9&pid=1-s2.0-S2405896315026518main.pdf (accessed June 8, 2022).

[36] Germany Trade & Invest. Industrie 4.0 – Germany Market Report and Outlook, May 8, 2018. https://www.gtai.de/en/invest/service/publications/industrie-4-0-germany-market-report-and-outlook-64602 (accessed June 9, 2022).

Modern technology automates not only production processes but also the organization of production and virtually the entire product cycle. Marketing research that defines the structure and volume of output is automated through artificial intelligence.

The actual production process in smart factories is also predicated on the operation of automatic machines coordinated and controlled by the Internet of Things. Logistics functions, production cooperation between firms and other B2B activities are also being automated. Finally, the interaction with the end consumer, the ordering, promotion, and delivery of goods, also increasingly relies on artificial intelligence systems. In fact, only the design, configuration and goal-setting functions remain in the hands of the individual.

A modern "smart industry" is emerging in which the *dramatically increasing role of the human mind goes hand in hand with the displacement of the human being from the direct participation in technological processes*. "Industry 4.0," based on the interaction with the Internet of Things, will be a prototype of such unmanned production, relying at the same time precisely on the power of human intelligence.[37]

So that an avalanche of human displacement from non-mediated material production does not create a mass of "superfluous people," for whom new jobs or decent living conditions adequate to the period in question have not been created in time, appropriate measures must be taken. As production evolves, these new places and these new conditions will arise in any case. The challenge is to ensure that there is no gap between, for example, the phasing out of dying out occupations and the increasing demand for new activities, leading to millions of "new vagrants" and "new beggars" for years or even decades, living on public handouts or being the object of harassment.

1.2.3 THE NEW INDUSTRIAL REVOLUTION AND SOCIAL DEVELOPMENT STRATEGY

The transition to a new technological mode and a fourth industrial revolution cannot in themselves be regarded as strategic priorities and goals of societal development. Production has never been and cannot be the main

---

[37] Boyes, H.; Hallaq, B.; Cunningham, J.; Watson, T.; The Industrial Internet of Things (IIoT): An Analysis Framework; *Computers in Industry* 2018, Vol. 101, 1–12.

interest of society; it has always been and is the means of its functioning. However, any social goal cannot be achieved without ensuring the development of production, which is why setting priorities and targets is one of the initial stages of strategy and the first to quantify future production, sales, employment, and other indicators that are necessary to understand the resources required to implement the strategy. In essence, the assessment of objectives is a gauge—a guide for all subsequent stages of strategy formulation, including the quantification of objectives that constitute the main content of programs to implement these objectives.[38]

The movement toward the sixth industrial mode, the new industrial revolution based on Industry 4.0, the Internet of Things, the control of technological processes through integrated sensors and the processing of large volumes of information are therefore a necessary part of the strategic development project.

The transition to a new stage of social production inevitably involves a profound change in social organization, which will have a major influence on the strategic goals that society sets for itself.

1.3 TRANSFORMATIONAL PROCESSES IN THE GLOBAL ECONOMY AND CIVILIZATIONAL DEVELOPMENT: OPPORTUNITIES AND RISKS

1.3.1 CHANGES IN TECHNOLOGY ARE ALTERING ALL ELEMENTS OF PRODUCTION AND THE SOCIAL FABRIC

Technological change leads to changes in all elements of the production process, including social relations of production and the type of society as a whole. The new industrial society (according to J. K. Galbraith Senior)[39] is left behind, as is the failed march into a "postindustrial future." A new, second-generation industrial society (NIS.2) is emerging to deal with the tensions of the preceding era. "The NIS.2 concept involves not only the development of a new industry as a mode of material production on a qualitatively new technological basis but also the transformation of social institutions into a new state of quality

---

[38] Bodrunov, S. D.; Convergence of Technologies—A New Basis for Integration of Production, Science and Education; *Economic Science of Modern Russia* 2018, No. 1 (80), 15.

[39] Kvint, V. *Strategy for the Global Market: Theory and Practical Applications.* Routledge, 2015, 103.

[...] The tendency for human economic activities to change, and indeed to die out, begin to emerge already at the stage of NIS.2."[40]

Due to this, there is a "need to look to the future, with a broader historical horizon, to find a way forward that blends the rationality of technicality with the spiritual wisdom of setting goals and objectives."[41] Production should not be subordinated to the pursuit of consumption volume or prestige, or even capital accumulation—it should be placed under the control of the human mind. But the human mind must undergo an evolution that changes the current hierarchy of values.

The development of material production technology draws us to the frontier where humans, for the first time, begin to disengage themselves from productive activities, from the direct extraction of their "daily bread." Material production, while changing qualitatively, retains its industrial form in technological terms and remains machine production. The fundamental distinction in the transition from the old industrial system to the new is the intellectualization of production, its knowledge-intensiveness, and the product's knowledge intensity.

The described "new-industrial" mode of production is knowledge-intensive in a way that overshadows material and human labor input, allowing man to virtually forgo using his own physical strength in the production process, remaining "inside" production, a participant in it, and therefore performing labor functions (which are becoming more and more intellectualized). The level of technology intellectualization in the "new industrial" mode of production would finally allow human beings to begin transcending the boundaries of production.

The new industrial society and economy of the twenty-first century should become the "negation of negation," the dialectical removal of the late industrial system, as presented in the famous work of J. K. Galbraith's titled *The New Industrial State* and the post-industrial information trends as examined by Daniel Bell and his followers.

What is the meaning of this "negation of negation"?

---

[40] Galbraith, J. K. *The New Industrial State*. Princeton University Press, 1967 (quoted from the original edition of the book, a copy of which was given to one of the authors of this book (S.D. Bodrunov) by J. K. Galbraith Jr in connection with the collaboration).

[41] Bodrunov, S. D.; Galbraith J. K.; *Concept of the New Industrial Society: History and Development*, Bodrunov, S. D. ed.; USEU, 2018, 76, 77.

It is rational and efficient not to focus on building beautiful utopias but to analyze the system of trends affecting material production development and assess the possibilities and consequences of their realization.

> *The new industrial society of the second generation (NIS.2)* is a society predicated on a new round of industrial material production development, marked by increased knowledge intensity, a transition to knowledge-intensive production, an acceleration of technological change and a transition to continuous innovation flow, and the completion of the integration of production, science, and education (including at the main production level).

What is needed is not only a technological leap but also the perfection of *all the components* of first-generation material production (materials, labor, production, application of knowledge, and the organization of production). Only then can we talk about entering a new industrial society of the second generation—NIS.2. This is why Russia, whose national economy has been undermined by the unprecedented deindustrialization of the post-Soviet period, needs to reindustrialize its economy on a new, high-tech basis, as one of the authors of this book has written repeatedly.[42]

The trend toward ever-increasing rates of technological change is fundamental at the new stage of development of industrial society (NIS.2). The "rapid acceleration" of technological development hallmarks the economic system of the coming society. The pace at which scientific advances are converted into industrial production and its components into an industrial product—industrial production acquires the character of continuous innovation—becomes fundamentally important. *Technology transfer*, an element of innovation, is no longer an occasional "innovation," but an integral part of modern, efficient production activities.

Changes in the technological basis of production in the transition to NIS.2 inevitably entail changes in the economic relations and *institutions* system due to the development of the new content and structure of social production. This economy, "assuming a revival in a new way of the features of the past, sets new challenges for the development of market self-regulation and private property on the one hand, and state influence on the economy on the other.

---

[42] Bodrunov, S. D.; *From ZOO to NOO: Man, Society and Production in the New Technological Revolution;* Voprosy philosofii 2018, No. 7, 112.

Indeed, the individualization, flexibility, and knowledge capacity of production, the widespread use of internet technologies in material production (and its continuing exchange), the increasing role of individual skills—all of these are driving forces for the development of small and medium-sized businesses and the need to develop economic freedom. In these conditions, the personal experience, energy, and talent of an entrepreneur–innovator are of paramount importance. In this regard, the new industrial economy of the twenty-first century is also the "negation of the negation" of the era of "classical" industrial capitalism and the beginning of late industrial capitalism, within which the formation of industrial empires unfolded."[43]

The new industrial economy of the twenty-first century is fundamentally different from that time. "Modern challenges condition development in many spheres of the public and state economic system. As a consequence, there arises problems establishing fundamental and applied science as one of the main branches of social production, as well as problems developing mass and publicly available professional and higher education in conjunction with the continuous professional development of workers. […] The development goals of complex integrated production units (PSE-clusters) and macroeconomic integration of production, science, and education, the problems of essential structural reorganization of modern economies, the tasks of displacing hypertrophied developed areas of intermediation make it necessary to use active state industrial policy and long-term investment public–private partnership. All other spheres of state regulation of the economy should be structured accordingly."[44]

The transition to mass creation and use of knowledge-intensive products also places significant demands on economic relations and institutions. "The synthetic nature of such a product brings about many changes in the system of economic relations and institutions. Particularly, ownership of

---

[43] Bodrunov, S. D.; *Formation of Russia's Reindustrialisation Strategy.* Institute for New Industrial Development (INID), 2013; Bodrunov, S. D.; Lopatin, V. N.; *Strategy and Policy of Reindustrialisation for Innovative Development of Russia.* Institute for New Industrial Development (INID), 2014; Bodrunov, S. D. *Formation of Russia's Reindustrialisation Strategy,* 2nd edition, revised and supplemented. In two parts. Part One. INID, 2015; Part Two. INID, 2015; Integration of Production, Science and Education and Reindustrialisation of the Russian Economy: Proceedings of the International Congress "Revival of Production, Science and Education in Russia: Challenges and Solutions." Bodrunov, S.D., ed.; LENAND, 2015.

[44] Bodrunov, S. D.; New Industrial Society: Structure and Content of Social Production, Economic Relations, Institutions; *Economic Revival of Russia* 2015, No. 4 (46), 20.

such a product encompasses a rights system embracing both the tangible object itself and its intellectual component."[45]

As to the costs of high-tech products, the costs of developing technology and protecting intellectual property rights are comparable to, and in some cases higher than, the costs of producing them. Hence, the paramount importance of intellectual property issues for the new industrial economy during the transition to NIS.2.

As S. D. Bodrunov states, "This society is really going to be new. First and foremost by the nature of socioeconomic relations. The new industry calls for a new face of the market and public regulation, as well as private enterprise and public ownership. Because of the fundamentally different, almost limitless availability/opportunity to satisfy nonsimulative human needs in NIS.2, the basic attitude of product appropriation—and the basic contradiction of capitalism between the public character of production and the private mode of appropriation—will fall sharply in importance." Production will become "detached" from the individual and "appropriation" will become an act of fulfilling a need as simple and as accessible as possible without affecting any other individual.

"This opportunity is arising as the technological advances of the industrial mode of production continue to unfold. Human beings, as the latest generations of technology develop, are not abandoning the industrial process, but putting a controlled and guided natural process at its core."[46]

When the technological basis of production changes, so do all of its other components: labor, products, and the organization of production. *But most important is that all these changes entail changes in economic relations, like property relations, inherent in this new generation of industrial society.*

1.3.2 *CHANGING OWNERSHIP AND ECONOMIC RELATIONS*

"Even at the present stage of society's development, even before the transition to NIS.2, one can see trends in the evolution of property relations leading to their socialization and dilution. Property relations (especially private property) were intended to confer on the owner an undisputed right to own, use and dispose of economic resources. Yet, the evolution

---

[45] *Ibid.*, 21.

[46] *Ibid.*

of economic relations has long led to the encumbrance of property with various burdens to ensure social and other responsibilities of the owner.

In this context, there are numerous landownership services that allow third parties to exercise, within certain limits, rights of use (rights of way, rights of access to water sources, rights of livestock grazing, rights of access to coastal areas, laying of communications, etc.). There are numerous restrictions and encumbrances on property rights relating to construction, transport and industrial activities relating to safety obligations, compliance with certain quality standards, environmental requirements, etc.

Particular attention should be paid to the evolution of intellectual property relations regulating the economic turnover of the most important resource of modern production—knowledge. Such phenomena as crowdsourcing, wikinomics, free software, open source, copyleft, etc., contribute to the development of free access regimes to intellectual resources. On the other hand, there is a fierce struggle for the "enclosure" of intellectual property.

This corresponds to two trends in the development of property relations that can be traced in today's economic system: 1. the conservation of existing relations; and 2. the dilution of the institution of ownership to the point of the so-called "abandonment of property."

The diffusion of the institution of property is taking place in various forms, obviously in the development of forms of co-ownership and use of property, as well as in the separation of the functions of ownership and use. The owner may temporarily give up the use of the property and transfer the right of use to another person: renting, leasing, co-working, various types of shared use (co-working, carsharing, kick sharing, timesharing, etc.). The sharing economy is already worth hundreds of billions of dollars a year, and its share is growing steadily and rapidly."[47] In China alone, the turnover of the sharing economy reached USD 1.05 trillion in 2019 and may reach USD 1.28 trillion in 2020.[48] The share of the sharing economy is thus close to 8% of China's GDP.

"The shift to temporary use of the property (without acquiring the right of disposal, and often ownership) is largely determined by the increased speed of technological change. It makes no economic sense to acquire full ownership of aggressors that will become obsolete in a few years. The

---

[47] Bodrunov, S. D. *Noonomy*, Cultural Revolution, 2018, 94–95.

[48] Bodrunov, S. D.; Noonomy: Ontological Theses; *Economic Revival of Russia* 2019, No. 4(62), 10–11.

owner of these units may assume further obligations toward the user in respect to repairs and modifications.

"Another trend that also leads to the dilution of ownership is the fragmentation of capital. It is not for naught that modern 'economic theory of property rights' pays so much attention to the problem of the splitting of powers and the dilution of property rights."[49]

The emergence of shareholder ownership leads to an even more complex split of ownership rights. "Shareholders no longer have full ownership of the capital. Moreover, the totality of their powers depends on the type of shares and the volume of their holding. The functions of appropriation within property relations have also evolved greatly: already in the first half of the twentieth century there was a splitting of these functions between the owners of capital and the managers."[50] Several researchers (Thorstein Veblen,[51] Adolphus Berle and Gardiner Means,[52] Stuart Chase,[53] and others) had already raised these issues before James Burnham became known as a "pioneer," coining the colorful term "managerial revolution" and arguing that capitalist society was being replaced by managerial society.[54]

"In fact, the split of ownership functions is even more profound than their division between shareholder and manager. J. K. Galbraith showed that the real use of capital is placed in the hands of an army of specialists who form the "techno-structure" of the corporation. But that's not all. After all, the ultimate user of the elements of capital is all employees, although each of them performs only a minor function."[55]

Moreover, in all these cases there is a kind of "stratification," or "splitting" of property along several lines: (1) assignment-ownership-administration usage; (2) distribution of each element of property rights

---

[49] China Sharing Economy Market to Exceed 9 trln Yuan: Report; *Xinhua,* November 2, 2019. http://www.xinhuanet.com/english/2019–11/02/c_ 138523206.htm (accessed June 9, 2022).

[50] Bodrunov, S. D. Noonomy; Ontological Theses; *Economic Revival of Russia* 2019, No. 4(62), 11.

[51] *Ibid.*

[52] Veblen, T. *The Engineers and the Price System,* 1921, Batoche Books, 2001. http://socserv2.mcmaster.ca/~econ/ugcm/ 3ll3/veblen/Engineers.pdf

[53] Berle, Adolf A.; Gardiner, C. Means. *The Modern Corporation and Private Property,* The Macmillan Company, 1932. http://www.unz.org/Pub/BerleAdolf-1932

[54] Chase, S. *A New Deal,* The Macmillan Company, 1932. (The title of this book, *The New Deal,* was used by F. D. Roosevelt for his campaign program).

[55] Burnham, J. *The Managerial Revolution. What is Happening in the World,* A John Day Book, 1941, 71.

between multiple actors in space and/or time and/or (3) by function (shareholder–manager–employee); and (4) by power authority.

The last aspect requires comment: power and property are correlated concepts. Ownership provides the ultimate power of the owner over the object of ownership (up to and including forms such as slave ownership). Accordingly, property right infringement tends to involve the use of volitional relationships—formal and legal or informal, up to and including criminal, violent acts. The above-mentioned forms lead to a gradual "unbundling" of power and, with it, to the removal, or "silencing," of the power aspects of social relations in the sphere of production in the broad sense of the word (in unity with exchange, distribution, and consumption).

Hence this conclusion: the significance of power as an institution will decrease/dissolve/dissipate, which is what happens in the historical process. Accordingly, the role of the state as the subject of power, the generalized *owner* of the rights for the development of society, will be gradually reduced.

An essential aspect of the property diffusion phenomenon (its "dissolution," or the decline in its importance to satisfy human need for goods in the transition to NIS.2) is the progressive redistribution of property benefits in favor of "non-property holders," or people who have no specific relationship to a specific property (tax restrictions, restriction of property), which confirms the trend toward the reduction of economic relations between members of society in favor of expanding noneconomic forms of their interaction.

Let us also note the direct impact of technological changes on property relations. According to S. D. Bodrunov, "Robots and artificial intelligence are taking the place of blue-collar and white-collar workers. What happens to property relations when several functions are transferred from humans to technetical beings? What about user liability, for example, if a robot driver causes an accident? The owner may be liable for damages.

And liability for traffic violations?

Usage and even management functions are slowly starting to "disappear" from the individual. Further evolution in this direction will only accelerate."[56]

These processes, along with the tendency to reduce the value of property ownership, lead to changes in the property system and the entire social

---

[56] Bodrunov, S. D.; Noonomy: Ontological Theses; *Economic Revival of Russia* 2019, No. 4(62), 11.

order. We can "predict with greater certainty that the sharing economy, the economy of split and blurred property rights, will prevail *at the NIS.2 stage*."[57]

Other researchers have reached similar conclusions. Thus, corresponding member of the Russian Academy of Sciences G. B.Kleiner observes that "we are seeing an expansion of rental relationships and benefit-sharing (leasing, sharing, co-working, co-living, etc.). Ownership rights are diluted, distributed in time and space between different actors, intertwined, forming a kind of 'carpet of authority'."[58]

Thus, the system of property relations in the transition to NIS.2 changes significantly, which "entails changes in the whole system of economic relations. The nature of the market is changing from spontaneous fluctuations in market conditions to the results of complex, coordinated actions by individuals with different and intertwined elements of ownership relations. The nature of state regulation is also changing—it shifts its focus on achieving consensus in the complex balance of economic interests resulting from the new nature of property relations and the new modification of market relations.[59]

1.3.3 GROWING TECHNOLOGICAL CAPABILITIES AND RISKS

In terms of understanding the new possibilities of modern technology, let us pay attention to a special feature of technology connectivity. The result of such a connection—*a kind of synergy of technologies*—can be different. In our judgement, it cannot be described in terms of known theories (e.g., wave theories) and leaves room for research.

> *Technology synergy* is an increase in technological effect that exceeds the sum of the effects of individual technologies when two or more technologies are combined.

Technology synergy, as well as the introduction of new technologies into an existing technological environment, is possible because of the *penetration* effect. Naturally, not every technology can be integrated with

---

[57] Bodrunov, S. D.; Noonomy: Ontological Theses; *Economic Revival of Russia* 2019, No. 4(62), 12.

[58] *Ibid.*

[59] Kleiner, G. B.; Intellectual Economy of the New Century: Post-Knowledge Economy, *Economic Revival of Russia*, 2020.

a new technological solution. The term *readiness* denotes the susceptibility of technologies to penetration.

Penetration is incorporating a new technological solution into other technologies and the various elements of the production processes based on them.

Readiness is the potential for the new technology to be absorbed by other technologies and the various elements of the production processes based on them.

It is through technology synergies that hybrid technologies, characteristic of the sixth technological mode (as discussed in the previous chapter), operate. The technologies of the sixth mode have high readiness interacting with each other, which determines the mass formation of hybrid technologies on their basis. It is precisely the penetration effect of technology that determines the formation of complex, holistic technical complexes that form technocenoses.[60]

Technocenosis is a community of technical products formed by analogy with biocenosis, which are characterized by technological interdependence and common purpose.

Both public consciousness and science are advancing toward the realization that the new technological mode not only has the potential to revolutionize individual and communal life, but can also realize its full potentialities if embedded in a new social mode. The President of the World Economic Forum in Davos has articulated this thought, "The more we ponder how to harness the enormous benefits of the technological revolution, the more we look closely at ourselves and the underlying social models that embody and create these technologies, the greater our capacity to shape this new revolution to make the world a better place."[61]

The technological prerequisites for a transition to a different way of meeting human needs and a different level of satisfaction are now emerging and, at the same time, the very mechanism for shaping those needs is evolving. This entails a lot of changes in social relations, institutions and ultimately in the social context that defines the vector of technological

---

[60] Bodrunov, S. D.; Noonomy: Ontological Theses; *Economic Revival of Russia* 2019, No. 4 (62), 12.

[61] The concept of "technotsenosis" was introduced by Boris Ivanovich Kudrin. See: Kudrin, B.I.; Studies of Technical Systems as a Community of Products – Technocenoses; System Research. Methodological Problems, Yearbook 1980, Nauka, 1981, 236–254.

The technological potential of society and a new understanding of human needs, which include not only material benefits but also intellectual ones, are altering the very orientation of social and economic development. "In essence, a development strategy is always socioeconomic in nature and is designed to focus on improving the material, spiritual and intellectual quality of life of the population."[62]

The changes in technology and social relations of the future are directly connected with the birth of a new human activity, which means a new kind of human.

"Humanity is at one of the most important junctures in its history [leading to either]:

- a turn to true intelligent human;
- or a dead-end road to a technotronic society where the elite meets ever-increasing and largely simulative needs, and the majority is engaged in service activities"[63]—with a possible loss of control over the development of the technosphere and the destruction of habitat.

Progress in technology not only holds potential positive prospects, but also, without a corresponding awareness of the risks of "misuse" of its results, significant dangers. "We are witnessing an outstripping development of the *technosphere* and lagging development of that part of social human consciousness which is 'responsible' for the intelligent use of technological advances and the sustainable formation of nonsimulative needs of individuals and society."[64]

The level of technological development that humanity permits can cause irreparable damage to civilization if there is no "balance" in the public mind to stop such a scenario from happening. In this sense, the current state of civilizational development can be described as a crisis, as many scholars have pointed out. For example, Russian Academy of Sciences academician B. N. Kuzyk believes that "the world is experiencing

---

[62] Schwab, K. *The Fourth Industrial Revolution.* Introduction, Penguin, 2017. https://www.litres.ru/klausshvab/chetvertaya-promyshlennaya-revoluciya-21240265/chitat-onlayn (accessed June 9, 2022).

[63] Kvint, V. L.; Theoretical Foundations and Methodology of Strategizing Kuzbass as the Most Important Industrial Region Of Russia; *Economy in Industry* 2020, No. 3.

[64] Bodrunov, S. D.; *From ZOO to NOO: Man, Society and Production in the New Technological Revolution;* Voprosy philosofii 2018, No. 7, 113.

a systemic crisis, above all a crisis of spiritual production, but at the same time there are crises of demographic, energy and environment, and food and technology. There is a shift in technological modes and a gradual transition to a new quality of life on a global scale. Such a 'parade of crises' imposes responsibilities on those who look to the future, develop and propose strategies for decision-making."[65]

The biological habitat of man is under threat; problems of interaction between man and the technosphere are accumulating; and the individual depends more and more on the technical and informational environment, which leads to the cyborgization of humans (so far without any significant physical invasion of the physical body). Humans *have become increasingly insecure in their existence as both biological and social beings.*

It is vital now to reduce the development of the most dangerous negative trends of modern civilization in its transition to NIS.2. "Two basic scenarios are likely here." One of them, conditionally "technocratic," is the way we firmly follow. This is related to the internationally accepted "economic development" paradigm, which refers to quantitative rather than qualitative progress.

But what is more important in satisfying needs: quantity or quality? If we have non-simultaneous needs in mind, it is purely quality. And this harmony cannot be verified by the algebra of statistical figures in the current "economy of crooked mirrors." As Joseph Stiglitz and Amartya Sen, Nobel laureates, noted in the aforementioned report of the Fitoussi Commission: "Expressing qualitative change is an enormous challenge, but it is crucial to measuring real income and real consumption, which are among the key factors of the material well-being of citizens."[66] And if we keep following the path of purely quantitative production increases, which we are currently following, it threatens to deplete resources, despite the availability of the latest technology.[67] Corresponding member of the Russian Academy of Sciences Konstantin K. Mikulski notes that there is

---

[65] Bodrunov, S. D.; Transition to the New Industrial Society of the Second Generation: General Cultural Dimension; *Economic Revival of Russia 2017*, No. 1 (51), 6.

[66] Kuzyk, B. N.; How to Successfully Implement the Strategy of Innovation Development of Russia; *Mir Rossii* 2009, No. 4, 5.

[67] Stiglitz, D.; Sen, A.; Fitoussi, J.-P. Mispricing Our Lives: Why Does GDP Not Make Sense? *Report of the Commission on the Measurement of Economic Performance and Social Progress*, Gaidar Institute Publishing House, 2016, 53.

a growing global awareness of the need to treat economic recovery as a qualitative rather than a quantitative growth task.[68]

The current economic system is gradually penetrating in NIS.2. But this stage in the development of economic society is of a transitional nature. The progress of the technologies of the sixth mode inevitably presents us with a choice: either the individual remains, changing the technological and socio-economic system, or the system changes the individual, or both change.

Obviously, both trends will come into play. But which one will become predominant? The individuals themselves with their own principles of communication, or self-development? The production of the material conditions of existence would then be left to *technical beings* (emerging from the upcoming Industry 4.0, artificial intelligence systems, etc.).

Technetical—referring to the techno-technological reality.
Technetics—the science of technical reality.

Needs that can be satisfied by purely technological means will not be the object of human activity. However, the definition of "technical tasks," social orientation, and goal-setting will be left to the individual.

But setting goals for the sphere of production directly depends on the values prevailing in society. This means that the values themselves must change suit. The cost of making a mistake in formulating goals in such an advanced technosphere, and a relatively autonomous one at that, would be immense. If the goals of such production are determined based on an outdated value system, both acute social conflict and conflict with the natural environment are inevitable.

Civilization may evolve in two ways: (1) as a technotronic civilization, i.e., the current human being will be de facto destroyed and replaced by other beings who will be able to exist in that environment; or (2) the human being may consciously create another trend, which is called "noociviliza-tion" in the theory of noonomy.

The mechanism of the first option is quite straightforward: we pursue a predatory course, "developing" the current "economy" by creating new simulative needs and satisfying them by producing more and more products (technetical, technogenetical species), i.e., through technological genetics. And then, these species will create a new environment themselves.

---

[68] Bodrunov, S. D.; Transition to the New Industrial Society of the Second Generation: General Cultural Dimension; *Economic Revival of Russia* 2017, No. 1 (51), 7.

Global Development Trends 31

Evidently, scientists pushing the boundaries of scientific knowledge are driven by good intentions: creating new medicines, correcting genetic abnormalities, etc. But they do not deny that these scientific advances could be misused, to the extent of creating new life forms or "editing" the biosubstance of human beings themselves.[69] How far will we go down this road? And what criteria will guide our decision-making? The choice of the further civilization's development trajectory depends exactly on the very answer to such and many other similar questions.

1.3.4 INCREASED ENVIRONMENTAL PRESSURE

The concept of noo-society has an undoubted connection with the idea of the noo-sphere by academician V. Vernadsky. His concept of the biosphere's transition to a noo-sphere can hardly be challenged in its rational form. "Vernadsky's central thesis—that since the twentieth century, humanity has become the leading geological force and is henceforth responsible for the reproduction of the Earth's biosphere—has been repeatedly confirmed by historical practice, in both positive and negative terms. Technogenesis[70]—the creation of the technosphere and filling it with techno-matter—already rivals biogenesis and the biosphere in terms of the mass of matter involved and energy expenditure."[71,72]

> *Technogenesis*—the creation of the technosphere and filling it with techno-matter and technetical beings.

The history of civilization shows us the accelerating growth of human-created technetical species, in strict accordance with the law of

---

[69] Mikulski, K.; On the Conceptual Elaboration of The Tasks of Modernization of The Russian Economy; *Society and Economy* 2010, No. 12.

[70] Bodrunov, S. D.; Noonomy: Ontological Theses; *Economic Revival of Russia* 2019, No. 4(62), 13.

[71] The term technogenesis was introduced by Academician Fersman. *See:* Fersman, A. E.; Geochemistry, V. 2., 1934, 27. *See also:* Balandin, R. K. Geological Activities of Mankind. Technogenesis, High School, 1978. For a definition of technogenesis, *see:* Kudrin, B.I.; Technogenesis; Globalistics: Encyclopedia; Mazur, I.; Chumakov, A. N., eds.; Center for Scientific and Cluster Programmes Dialogue, Raduga Publishing House, 2003, 998.

[72] For a large body of data on anthropogenic pressures on the biosphere, *see:* Karlovich, I. A.; Regularities of Technogenesis Development in The Geographical Envelope Structure and Its Geo-Ecological Consequences; *Specialty,* Vladimir, 2004.

the "accelerating acceleration" of innovation, to the detriment of the rapidly increasing diversity of biota. The resultant increased pressure on the habitat, due to the simulated growth of human needs and the increased use of natural resources to satisfy them, together with the expansion of resource extraction and processing areas, pose a real threat of negative (catastrophic!) consequences for civilization.

These are the factors in the crisis scenario of civilization:

- immensely growing and predominantly simulative needs;
- risking loss of control over technological development driven by the pursuit of artificially inflated needs;
- extremely high level of technological development that allows irreparable damage to civilization;
- increasing human dependence on the technical and informational environment;
- accelerating growth of human-created technetical species to the detriment of the rapidly displacing diversity of biota;
- increasing technological pressure on the habitat;
- advancing the technosphere while the part of human society responsible for the intelligent use of technological advances is lagging behind;
- weakening internal regulators of intelligent behavior determined by the content and level of cultural development.

In terms of resources, a shift in priority from traditional (material) resources to the basic NIS.2 resource—knowledge embodied in technology—should be definitively prioritized. And in gnoseological terms, a change in priorities and in the development focus itself is necessary.

This is well illustrated by data characterizing the current state of our civilization's environment, which has been created according to modern "economic growth" paradigm trends. Here is, for example, the total amount of what humanity has done in the 5000 years of our existence: according to geologists, the *weight of the technosphere*, i.e., everything that humanity has created in its history through technology, *is 30 trillion tons* (Table 1.1).

TABLE 1.1 The Approximate Mass of Main Components in the Physical Technosphere (in Descending Order, 1 Tt = 1012 tons)

Component	Area, 106 km²	Thickness, cm	Density, g/cm³	Mass, Tt	%
Urban areas	3.70	200	1.50	11.10	36.9
Rural housing	4.20	100	1.50	6.30	20.9
Pastures	33.50	10	1.50	5.03	16.7
Cropland	16.70	15	1.50	3.76	12.5
Trawl seabed	15.00	10	1.50	2.25	7.5
Land use and soil erosion	5.30	10	1.50	0.80	2.7
Non-urban roads	0.50	50	1.50	0.38	1.3
Afforestation	2.70	10	1.00	0.27	0.9
Water bodies	0.20	100	1.00	0.20	0.7
Railway tracks	0.03	50	1.50	0.02	0.1
Total (if applicable)	81.83			30.11	

Source: Scale and diversity of the physical technosphere: A geological perspective. Zalasiewicz, J.; Williams, M.; Waters, C. N.; *The Anthropocene Review* 2017, *Vol. 4(1),* 12.

At this point, humans have already transformed so much mineral, non-living nature that they have created far more in the last 500 years than nature (non-biological "civilization") has transformed in hundreds of millions of years. That is, according to other specialists (also geologists), we can talk about the onset of a new geological era. They call it the Anthropocene.[73] Yet geologists describe it from an external point of view; the author's views are based on what's inside it, what it is made of—our irrational, albeit highly scientific, use of technology.

Another estimate: according to biologists, *over the 4.5 billion years of the Earth's existence, the weight of the biota* (i.e., what nature has created) *is about 2.5 trillion tons*. That is, we have created 12 times more in a few thousand years (and mostly in the last 100 years) than nature has in billions of years. Are these not signs of major changes, right up to the onset of the crisis mentioned above? The species diversity of biota is variously estimated at between 8 and 100 million species, and the species diversity of technical species (human-made products of all kinds) already exceeds that by about a thousand times. And according to some estimates,

---

[73] Bodrunov, S. D;. From ZOO to NOO: Man, Society and Production in the New Technological Revolution; *Voprosy filosofii* 2018, No. 7, 73. *See:* Issberner, L.R.; Lena, F.; Anthropocene: Scientific Debates, Real Threats; *UNESCO Courier* 2018, No. 2.

we increase the number of such species by an order of magnitude every 10 years or so!

The renowned international organization Global Footprint Network (GFN) has proposed a sound methodology for calculating the so-called ecological debt and each year sets out based on this methodology:

"Ecological Debt Day" is the date when the amount of resources consumed by humanity exceeds the number of resources the Earth can recover in a year. In 1970, such a day was in December, i.e., there was no ecological debt. Since the 1980s (the beginning of the globalization period; you can see a clear correlation!), it has emerged and continues to grow. In 2019, Eco Debt Day was around July 30.[74]

In 1970, we consumed 0.9 of the resources that nature can regenerate in 1 year; in 2021, this figure is 1.7.[75] The growth rate has doubled and is accelerating continuously. GFN extrapolation shows that at this rate, our eco-debt will be over 400 years by 2050.

Apparently, the direct pursuit of such a developmental trajectory drives our civilization to irreversible negative consequences. What is needed, then, is a strategy that changes the fundamental paradigm of our development.

Another important consequence of current trends in society is the negative impact on human nature itself.

1.3.5 THE DANGERS OF INTERFERENCE WITH HUMAN NATURE

The potency of the contradictions involved in interfering not only with the external environment but also with human nature itself is rooted in the very progress of technology. For example, information and communication technologies (ICT) and artificial intelligence (AI) technologies offer new opportunities for human interaction. A large part of communication has already been moved to the virtual space of computer networks. People do not interact directly there, but their virtual imprints, virtual clones ("avatars," profiles, accounts, etc.), sometimes radically different from their real prototypes.

---

[74] "The fact that has shocked me the most is the Overshoot Day: By July 29th, we used up all the regenerative resources of 2019. From July 30, we started to consume more resources than the planet can regenerate in a year. It's very serious. It's a global emergency." (Pope Francis, La Stampa, August 9, 2019). *Earth Overshoot Day* 2019. https://www.overshootday.org.

[75] Garrett, C. Earth Overshoot Day 2022: What is Earth Overshoot Day? Climate Consulting by Selectra. https://climate.selectra.com/en/environment/earth-overshoot-day (accessed June 9, 2022).

Is that a good thing or a bad thing?

Ethical evaluation ("good or bad" or "good or evil") is more than apt here. We are addressing moral issues of the world, where people can thus solve creative information-cognitive tasks by transplanting all sorts of routine and secondary functions to virtual personas. When equipped with AI systems, such virtual identities can take over, for example, the accumulation, processing, and sorting of information flows. A self-learning AI can absorb new knowledge and even apply it to new objects, but it cannot discover previously unknown knowledge. So, for the time being, there is no need to be wary of competing with humans as a species on this side (which is not the case with the individual human professions).

But who will use this virtual world, how, and for what purpose? What will be the rules, what goals will be served by communication in virtual space?

The technosphere has become a colossal and largely human-independent force heightening our responsibility to put tit within a reasonable framework that precludes the spontaneous destructive impact of technological processes. This responsibility may be realized and transformed into a system of collective action. Or it may not be realized, or realized but not implemented due to the collective irresponsibility of humanity. There must be a line. Why? Because the process and result of needs satisfaction also change human beings themselves, their physical and intellectual properties. Needs, without a rational distinction between real and simulated, cannot only fundamentally change the individual properties of an individual as a being with both natural opportunities and limitations to consumption but can also change his or her very nature. The technological prerequisites for such developments are being created right now. For example, the Massachusetts Institute of Technology (USA) is already editing genes inside the human embryo, removing (disabling) some and adding others. And another American institute (The Scripps Research Institute, TSRI) went even further: to the four nitrogenous bases that make up DNA in nature—adenine, thymine, guanine, and cytosine (of which every living thing is built, from a bacterium to a whale). Researchers added two artificial bases that do not naturally exist within us, inserted these aliens into the DNA of living cells, and successfully forced them to reproduce, with the acquired (embedded) properties being inherited and producing semi-synthetic proteins.[76] As Bodrunov states, "But if people want to change their nature, then what are we talking about: a person as a biosocial being—or another creature? If we are talking about human beings, we assume that there are reasonable constraints that do not allow for such a development."[77]

A new type of human being is emerging with the development of a new type of production, a previously unprecedented level of knowledge intensity with the growth of technological power and a tremendous capacity to meet needs. What will it be? The question is by no means a foregone one. And already now, different variants of human development within the new industrial civilization are visible.

The current state of technogenesis leads human beings into a perplexing and poorly managed technospheric world that evolves according to its own laws. A society based on production relations that prioritizes profit and other volumetric value aggregates (e.g., GDP) as production goals are not inclined to accept the risks and threats posed by the subordination of technology to the extraction of profit.

Will humans be able to meet the challenges of this new, technotronic or technogenetic civilization? Will it lead to a society of humanism and the spread of knowledge-intensive human activity; a society in harmony with nature and overcoming social conflict, where an individual is preoccupied with the appropriation of knowledge; a society where material limitations play a minor role since along with access to material needs, the private appropriation of material goods will lose their dominant position? Or is the opposite waiting for us?

In developed countries, people, overwhelmed by the nearly limitless possibilities to increase their needs, may be tempted to over-consume. In less-developed countries, the former chronic under-consumption of billions of people threatens to turn new technological possibilities into an unrestrained quantitative growth of material goods beyond rational limits. Both tendencies are fraught with inflating irrational, fictitious, simulated needs.

---

[76] Medvedev, Yu; Six-letter life. The First Bacterium with Synthetic DNA Has Been Created; Rossiyskaya Gazeta – Federal Edition 2017, No. 7448 (282). For more details *see:* An Organism with DNA Containing 6 "Letters" Has Been Created; XXII Century: Discoveries, Expectations, Threats, *Popular Science Portal,* January, 2017. https://22century.ru/biology-and-biotechnology/42655 (accessed June 9, 2022).

[77] Bodrunov, S. D.; Noonomy: Ontological Theses; *Economic Revival of Russia* 2019, No. 4(62), 13.

The type of human consumer, in the perpetual pursuit of fictitious goods, with no regard for anything, is becoming widespread. Pressures on the Earth's resources will increase despite the potential for significant reductions in the intensity of production. Unbridled consumerism threatens to consume any amount of natural resources and overwhelm the Earth with waste, or even to plunge humanity into conflict over material goods and the scarce resources for their production.

What emerges is a world of estranged people—estranged from others, from society, and from their own being, after all. Humans are dehumanized and become human *as it were*, threatening the existence of their environment and themselves. Many people on Earth are caught up in the vortex of mindless pursuit of fictional consumption growth that eats into real resources and people themselves.

Can this path to gridlock be avoided?

1.3.6 INCREASED RISKS ARE INEVITABLE WITHIN THE EXISTING ECONOMIC SYSTEM

Society is not yet "mature" enough to make proper use of technological progress and its achievements. It has not matured, partly because technological advances have not yet "fed" everyone. Why is it that today, when the world produces enough grain to feed everyone, millions go hungry? Because there is still a so-called capitalist (let's say economic) way of appropriating these very goods. In the current model of satisfying people's needs, the essence of which is economics, technological progress (facilitated by financial capital, which absorbs its results) enables income to be redistributed in favor of financial rather than productive capital, not in favor of satisfying people's real needs.

Financial capital is ready to profit at the expense of anything—at the expense of other people, nations, countries, etc., at the expense of innovative margins, at the expense of new and pseudo markets, at the expense of the simulated orientation of the consumer toward what is supposed to be bought. And every time, it transfers useful resources to increase financial profit without increasing the real product.

This situation arises at every stage of technological progress. The transition to each new technological mode, as world history shows, was very often accompanied by expansion, wars, conflicts, etc. While meeting needs should allow people to live better lives.

Why is this happening? Because there is disharmony, a lag of social consciousness behind the possibilities of technological progress.

Why is the situation more acute today than before? Because technological progress in any new stage always gives much more opportunities than in the previous one. If they are used incorrectly, the risks increase dramatically. The technological advances have achieved such a level and capability that almost any terrorist can build an atomic bomb. A society that is not mature enough to benefit from the fruits of science and technology in this way is a threat to itself.

On the one hand, the socioeconomic system is very coherent and, on the other hand, dynamically evolving. The interconnected elements of the system, as they develop, influence one another and they develop with different "velocities." Disharmony, velocity dysfunction, and non-alignment of the velocities of system elements development can break the system, as the stress of the connections cannot be infinite.

Every transition and technological change led to a change in technological modes. The production modes have each time shaped a new type of society: the industrialized mode of production—the new technologies of that phase led to the formation of capitalist society, not the other way round. Each new stage now offers new, much broader opportunities to meet human needs. But if these needs are not reasonable, then we will use technological progress as a tool that was given to a child or an underdeveloped being, so to speak. Now humankind is once again in such a situation, and the opportunities of the current stage are so gigantic that if misused, they can lead us immediately to the brink of disaster.

At a certain point in time, the development of commodity markets led to the birth of financial capital to serve them, making money the "lord," the suzerain of economic relations. And then money and financial markets, by their very nature and their need to constantly expand their scope and capture areas of influence, began to influence the structure and infrastructure of so-called international trade in a decisive way. Corporations, having captured national markets, have gradually expanded beyond national territories. The emergence of multi-industry conglomerates and transnational business structures, their "intertwining," "sanctioned" by capital spillovers, formed the basis for the global marketplace.

Financial capital itself now dictates to political forces the rules of policy making in all spheres of social life. Hence, all kinds of alliances (in the first sense, trade and economic), trade wars, pseudodemocratic sanctions, etc.

Often, awareness of the pitfalls of the current globalist economic model requires national and cross-country countermeasures.

"The key role in countering the challenges of global financial markets," writes RAS Corresponding Member Mikhail Golovnin, "is the construction of its own system of protective mechanisms as a flexible set of measures related to the introduction of certain restrictions on cross-border capital flows and macro-prudential policy measures. Measures aimed at accelerating economic growth and restructuring the economy should play a central role in this system."[78]

The process of globalization relates to the technological development of human society and civilization. Moreover, it is predestined by it in the paradigm of "zoo-development" in some sense. Why? Because by providing capital with expansion opportunities, under the conditions of "under-acculturation" of humans in general (under-"noo" in the sense of limiting simulative needs), technological progress is put at the service of little more than limited financial capital, allowing it to overflow more effectively, to be used, etc., while satisfying the possibility of capital, the desire of capital, and of the capitalist to multiply capital.

Today, the situation regarding the formation and satisfaction of simulation needs has gotten to a tipping point. Technological progress, being an instrument of finance capital, creates new and more simulative needs and immediately satisfies them, drawing everyone and everything into the process: on the one hand, the product must be available somewhere, on the other hand, it has to be possible to buy the product. But, on the third side, it uses untapped resources to promote this same product as effectively as possible and to build up financial capital.

Historically, we are approaching a point where the global financialization process has already taken over all major venues. There is no longer any room for dramatic territorial acquisitions: no room for extensive expansion; there is no other way to expand except to intensify the process—exploiting nature, especially natural materials, raw materials etc., without regard to damage.

The perilous thing about this development is that it does not stop there; it has, as we have already noted, a negative impact on the formation of a personality. Why? Because it "inadvertently" creates needs in people that they have not thought of before and which are in fact simulative, and thereby harms positive personal development. Financial capital also

---

[78] Golovnin, M. Yu.; Challenges for Russia's Economy from Global Financial Markets; *VEO Russia Proceedings* 2019, V. 218, No. 4, 319.

advances in the social sphere, changing people's attitudes, artificially creating consumerism, encouraging people to absorb mass-cultural phenomena, etc., which are not important for human beings.

The link between technological advancement and financial capital, globalization through financial capital, has a consequence—it dictates the conditions for society to exist. Accordingly, the promotional structure of financial capital, etc., is formed. One could argue, for example, against the creation of new rules for international trade, precisely because modern international trade in its present form is a mechanism for promoting financial capital interests and imposing (preferentially) simulative needs.

Every current product is the result of recycling tons of natural matter. For example, a pair of shoes requires 10–30 tons of fresh water in production, while many of them are disposed of without being sold or are used only occasionally. We could claim that this way of meeting human needs is too wasteful and cannot be sustained indefinitely.

Another prime example is Cambodia. Introduced to the jungle country by capitalist colonizers a hundred years ago, Hevea trees have grown into a mighty plantation. Hevea trees yield sap for the rubber industry for 20–30 years; after that the tree becomes useless, and the plantation becomes a graveyard of deadwood; the surrounding jungle must be cut down every year, new trees are planted, and the plantation grows. The United Nations predicts that by 2030, Cambodia will be a country with no natural forest at all. It is transformed into a land of casinos, banks, shady capital, pimping, etc., with natives migrating or joining the ranks of the poor in the fast-growing cities (in terms of growth and urbanization as well as GDP (!), the country has for many years formally overtaken even China).

This is how—objectively, by virtue of its nature—financial capital operates. Destroying not only natural phenomena, flora, and fauna, but also human beings and society. There is a deep connection between this destruction and financial capital.

1.3.7 STRATEGIC MILESTONES AND STRATEGIC RISKS

The study of technological trends lets us understand the impact of technology on the social fabric. The technological and industrial revolution is inevitably followed by a qualitative shift, placing the development of society at the stage of a new industrial society of the second generation (NIS.2).

Can reaching this milestone be seen as a strategic development goal? Yes, if we are not talking about "labels" but about the substantive changes that characterize NIS.2. These are the goals of society. The set of elements that form these objectives include:

- increasing knowledge-intensive production, which presupposes a fundamentally increased role for science and education and their close integration with industry;
- reducing the share of material costs and increasing the role of knowledge in the final product;
- developing workers capable of mastering new knowledge and applying it in production;
- an increasing level of needs saturation with a significant reduction in the cost of their satisfaction;
- changing the system of economic relations toward the blurring of property relations, the development of forms of direct access to goods, and their joint appropriation.

However, the transition to NIS.2 is impossible without achieving the intermediate goal—the reindustrialization of production. This is not about shutting down services. To take the economy to the next level, a new quality of material industrial production is needed—only on this basis can the rest of the economy be technologically transformed. For Russia, this intermediate goal is all the more important as we need to address the consequences of the deep deindustrialization of production in the 1990s, which led to the degradation of the national scientific and technological core of the economy and weakened domestic drivers of economic development.

Herewith we should avoid a technocratic approach that relies on the automaticity of societal progress and technological progress. Uncontrolled and unguided technological advances and growth in production bring risks that are already out of control. As Samir Amin, one of the founders of world-systems thinking, remarked, "In an age like ours, when there are enough weapons to destroy the entire Earth, when the media can tame crowds with frightening effectiveness, when short-term egoism or anti-human individualism is a fundamental value that threatens the Earth's ecological survival, barbarism can be fatal."[79]

---

[79] Amin, S. Russia and the Long Transition from Capitalism to Socialism. *Monthly Review Press,* 2016. *See also:* Amin, S. *Russia: the Long Way from Capitalism to Socialism,* Bodrunov, S.D., ed.; INID, Cultural Revolution, 2017.

In fact, society is approaching a civilizational crossroads that puts us in front of the choice described above.

Will we allow continued thoughtless interference in the natural environment, depleting natural resources, and destroying the natural balance? Can we allow the deployment of mindless interference in human nature, disguised by the pursuit of whatever needs we may have? It is probably necessary to ensure that such development trends are reduced, as these risks have already started to materialize.

Development strategy must provide an answer, including how to avoid the increasing threats to human civilization.

1.4 NOONOMY AS A CONCEPTUAL PLATFORM FOR THE GLOBAL TRANSFORMATION OF SOCIETY

1.4.1 THE NOO-APPROACH AS A CONDITION FOR OVERCOMING CIVILIZATIONAL RISKS

If we don't add other knowledge—about the importance of conscious self-restraint, new approaches to the organization of our lives, and above all the possibilities of technological progress—to our growing knowledge of technology we will obviously face a catastrophe. Ahead is the singularity point of our civilizational development. We may pass it unnoticed, but the consequences will be felt very soon. The choice is either to go on with "zoo-life" with "zoo-economy," "zoonomy," and then what is said above awaits us, or to go out into NIS.2 and gradually form a world of intelligent needs and intelligent production, noo-needs, and noo-production.

Technological breakthroughs into the future will only take humanity a real step forward if they are based on radically new approaches that alone can point us in the right direction toward harnessing our increasing—and therefore potentially dangerous, yet very lucrative—technological potential.

The noo-approach *means connecting technological power with the power of knowledge, with human reason, embodied in the traditions of human culture.* The cultural codes of society are now the sine qua non-condition for the technological use of knowledge, *and our cultural norms determine what will become of our times.*

The new technological possibilities, which provide the basis for the human factor to emerge from direct production, form the basis for the withering away of economic relations (competition over the use and appropriation of resources and the results of production). But society itself will undergo profound changes as a result. The answer to the challenges of the extensive "technocratic" development scenario, which leads to a deadlock of civilizational crisis, should be a conscious intensification of the creation and use of technology, contributing to the personal development of humans to improve the cultural code of modern civilization. As Bodrunov states, "Public institutions will also change because of the widespread, "pervasive" application of such technologies. For example, real direct democracy becomes possible—not only (and not fully) regarding elections, but also the direct resolution of any community issues based on a consensus of trust (not requiring verification)—whether to put a tram in the street, demolish a monument, build a factory next to a residential area, etc."[80]

We emphasize that technology development in this option would aim to achieve "reasonable societal development and meet the individual's reasonable (nonsimulative) needs within the framework of the cultural-civilizational code formed on the NIS.2 basis. No matter who is working [whether] a robot or a human creator. The basis will remain material, and the mode of goods production will remain industrial, based on the technology of the time."[81] More precisely, it will remain noo-industrial to meet the needs of a noo-industrial society existing in the noo-sphere.

On the face of it, these theses follow from Vladimir Vernadsky. However, if you read more deeply, it's something else. Many thinkers of the past (Karl Marx, Vladimir Vernadsky, Erich Fromm, the Club of Rome theorists, etc.) appealed to human reason to solve increasing problems. However, there was no answer as to what specific material means we could achieve with such a solution, nor was there an answer on how to resolve the contradictions that had arisen."[82] It seems to us that we can now answer this question: From a purely humanistic interpretation of the noosphere, based on socio-philosophical reasoning, one must move on to

---

[80] Bodrunov, S. D. Transition to the New Industrial Society of the Second Generation: General Cultural Dimension; *Economic Revival of Russia* 2017, No. 1 (51), 9.

[81] *Ibid.*

[82] *Ibid.*

the understanding that "these ideas can be implemented on a solid foundation of material production trends."[83]

In this sense, the rationale for the concept of NIS.2 and the theory of noonomy turns into an approach to justify a new stage in the development of human civilization. We would propose to call it a noo-civilization in which production would be not so much a realm of technology as a realm of human reason (based on the purely material processes of noo-industrial production, outside of which humans could neither sustain their own existence nor evolve).

Simultaneously, the social role of knowledge as a means to discover new, more efficient and economical ways of satisfying rational human needs (as opposed to the current quantitative increase in consumption, which has visible limits) and as a solution to the tensions and contradictions that accompany profound technological and societal shifts will increase dramatically.

But technology is not in itself the creator of a new society in which the key role is played by knowledgeable, intelligent human beings (and this is our fundamental contradiction to the technological determinists).

It is culture (morality, so-called basic values, etc.) that "is the means for the formation of the most important element of the civilizational code of such a society—the internal self-restraint of the individual, which redirects it from the unrestrained increase in consumption and the pursuit of all kinds of mirage-simulacrums to the formation of the needs of a reasonable person, where the quality of needs and consumed goods is of paramount importance."[83] It is also "the basis of a new quality of interpersonal interaction, both in the creative work process and social life. At the same time, advances in technology offer enormous potential for changing the very cultural code of human civilization.[84]

The question of what social arrangements would enable us

to set goals for production and technology relevant to human development, to guide the development of technology so that it fulfills that goal, is at the heart of the evolution of the social order in the transition to a new society.

The development of NIS.2 in its transition to noo-civilization will unequivocally lead first to a change in the standard role of society's basic,

---

[83] *Ibid.*, 10.

[84] *Ibid.*, 9.

familiar institutions—the state, money, the means of appropriation of social wealth—and then to their gradual disappearance. A stable state of social structure will come about based not just on trust but on the firm knowledge that the information resulting from the "public" exchange is always true. Knowledge may be different, but the need will increase for the *right* kind—tested, trustworthy and reasonable.

The role of the mind is increasing in leaps and bounds, and it matters what that mind becomes. Will it rely on the cooperation of people to achieve high goals, or will it be given free rein to the darker side of the power that is in knowledge? Educating the rational (and also cultural) human being becomes imperative in shaping the society of the future – as does the question of how people will be able to cooperate and solidarize to achieve common goals.

1.4.2 REJECTING ECONOMIC RATIONALITY: WHAT INSTEAD?

As the technological possibilities of satisfying intangible/cultural/spiritual needs increase, human society reacts appropriately to this by changing the trend of civilization—primarily the values and their bearers, as well as their behavior accordingly. For what discovery did Richard Thaler receive the Nobel Prize in Economics in October 2017? For confirming that people (especially young people) are increasingly guided in their economic behavior not by "rational" considerations but by their emotions.

Emotions are spiritual, intangible elements representing the cultural value component of the average person's overall needs structure. And humans have always been governed by satisfying this component of needs, which is not always believed by dry economic rationality. Significantly more advanced in this respect, Generation Z increases the overall proportion of this kind of need (emotionally colored) in the overall structure of societal needs.

Hence, the perceived growth of increasingly less "rational" (within the relevant rationality framework) decisions by "market actors" from the perspective of apologists for the influence of human biosociality on the social order. A considerable number of market "generals" and "strategists" have not yet realized that the market is gradually becoming, in a sense, a relic of a bygone past, of a previous economy; that the increasing trends of such "irrationality" are just "sensors" recording the increasing change in human preferences and the declining importance of "rational market" behavior as well as the market itself for human beings.

The mentioned work of R. Thaler shows what economists began to recognize: an individual in his or her life is not guided by "indifference curves" from the economics textbook. In most cases, he or she makes decisions based on various criteria, including nonmarket ones. And the goals of production and leading needs have always been and are shaped by nonmarket means, even in the most market-driven and capitalist world.

"With the development of noo-society, with the transition to no-production and no-necessities, there is a transition from economic rationality to no-rationality, and this new character of rationality and consequently the new determination of development goals take on paramount importance and serve as the basis for changing the character of social relations which are becoming increasingly noneconomic."[85] *Noonomy* thus replaces economy. It relies on a shift from a growth paradigm based on economic "rationality," focused on increasing volumetric values, to a development paradigm based on achieving concrete goals, satisfying real human needs.

> *Noonomy* (primary definition) is a noneconomic way to organize economic activity, oriented toward meeting specific human needs based on the criteria of reasonableness governed by the development of knowledge and culture.

The concept of noonomy echoes, both terminologically and semantically, the idea of the noo-sphere. The idea of the noo-sphere was first proposed by Édouard Le Roy (1870–1954), Pierre Teilhard de Chardin (1881–1955), and Vladimir Ivanovich Vernadsky (1863–1945). The impetus for developing these ideas developed from Vernadsky's 1922–1923 Sorbonne lectures on geochemistry attended by Édouard Le Roy and Pierre Teilhard de Chardin. The term "noo-sphere" was first introduced by Édouard Le Roy.[86] A detailed interpretation of the noo-sphere was presented in the works of Pierre Teilhard de Chardin and Vernadsky in the late 1930s.

As understood by Pierre Teilhard de Chardin, the noo-sphere represented a qualitatively new state of consciousness concentration, which formed a special sphere of the spirit, a "thinking layer" covering the Earth.

---

[85] *Ibid.*, 10.

[86] Bodrunov, S. D. Noonomy: Ontological Theses; *Economic Revival of Russia* 2019, No. 4(62), 14.

The concentration of thinking on a planetary scale is closely aligned with the fusion of the human spirit, which will lead to the emergence of the "Earth Spirit" as a result of further evolution.[87]

V. Teilhard de Chardin Vernadsky, in turn, approached the idea from the perspective of natural science, pointing out that intelligent human activity becomes the main transforming force concerning both the biosphere and the geological shell of the Earth (the biogeosphere).[88]

What we see in all these concepts, however, is not a scientific theory, but rather an interpretation of the undeniable fact that the activity of humans and human society, with their inherent capacity for intelligent action, is becoming a determinant in the state and evolution of the Earth itself (at least for now), the Earth's surface, and a major determinant of the fate of humanity itself.

The primacy of reason inevitably raises a development problem, understanding what imperatives *for it* will become predominant. This provokes the question of how human society should be structured to predetermine the intelligent use of a tool as powerful as reason, so that reason is not just used as an efficient tool to satisfy the zoological instincts that have been perverted by modern civilization. There is no answer to this question in the idea of the noo-sphere.

This answer is given by the theory of transition to a nonsocial order, to a *noo-society*. And *noonomy* is one of the basic elements of noo-society as a kind of planet-wide "nomos" (law, order, etc.), defining a noneconomic mode of economic activity and satisfaction of human needs, oriented toward cultural imperatives rather than economic rationality.

The term noonomy is derived from the Greek words "noos" (νους—reason) and "nomos" (νομός—order, law). You would think that since noonomy is defined as a mode of economic activity, why not use the Greek word "oikos" (οἶκος—home, household) for such a term as well. However, in the modern scientific tradition, terms derived from this word are used to refer to economic reality. Thus, noonomy avoids its identification with a particular economic structure of society.

"We do not proceed from a mechanical combination of the terms 'noo-sphere' and 'economy' but from an understanding of the Greek term 'noos' as reason, which is based on the criterion of truth as a conscious, timeless

---

[87] Le Roy, E. *L'exigence Idéaliste et Le Fait De L'évolution*, Boivin & Cie, 1927.

[88] *See:* Novikov, Y. Y. ; Rezhabek, V.G.; *Contribution of E. Le Roy and P. Teilhard de Chardin in the Development of the Noo-sphere Concept.* http://www.nffedorov.ru/w/images/3/36/Lerua.pdf

value. As early as the 11th century, Metropolitan Hilarion in *The Word of the Law and Grace* wrote: "He has led us to the true mind."[89] In this sense, to reduce the Greek word "noos" to its Latin counterpart "razio" is deeply mistaken.

Rational is that which meets certain criteria—but are the criteria themselves reasonable? The economy is always rational, but do economic actors always act rationally? And are they able to go beyond the criteria imposed on them by this economic system?

Noonomy presupposes a different way to evaluate economic action, a different way to assess needs—one based not on rationality but reason, on the "noo" that comes from an understanding of the true consequences of economic decisions and the true value of the needs that are met. It is thus not about economics, not about the rationally pleasure-maximizing individual, but about a different way of forming and satisfying needs, which can be called *noo-needs*.

On the other hand, the other half of the term noonomy, "nomos," is an ancient notion which, in the philosophy of the first third of the twentieth century, was applied to refer to a basic principle of organization of any space,[90] an absolute law of the existence of all things. That is to say, "noonomy" is an established mode, a way of meeting needs in a society where there is 'light of reason' and no relationship to production and relations of production; no relationship to property and relations of ownership; no economy and no economy is possible. *It is a noneconomic way of satisfying noo-needs.* Therefore, it is wrong to speak of a 'noo-sphere economy'—it is like speaking of a noneconomic economy, a non-predatory predator, etc."[91]

In a market economy, rationality is understood only as the maximization of monetary income. Obviously, the neoclassical economic theory argues that it does not reduce the issue to money, that it is inherent in human beings to maximize the receipt of any goods—but they are only really considered when they receive a monetary value.

It is only relatively recently, under pressure from behavioral economics research, that "neoclassics" has softened its stance somewhat, admitting that humans are not a programmed calculator of gains and losses, that

---

[89] Vernadsky, V. Scientific Thought as a Planetary Phenomenon, *Nauka*, 1991.

[90] Word on the Law and Grace, Platonov, O.A., ed., *Institute of Russian Civilization*, 2011, 70.

[91] *See:* Schmitt, K. *Nomos of the Earth in the Law of Nations Jus Publicum Europaeum,* Vladimir Dal, 2008.

other motives can drive them and that their economic decisions can be influenced by noneconomic factors as well. However, all this has been interpreted as human "bounded rationality": true rationality is still seen as counting gains and losses, but humans are supposedly imperfect. Various extraneous factors limit their capacity for rational behavior.

Actually, this is largely (though not entirely) true for a market capitalist economy because this is the criterion base of capitalist rationality. However, changes in social conditions of production entail changes in the criteria of the rationality of human behavior (Fig. 1.2). With the transition

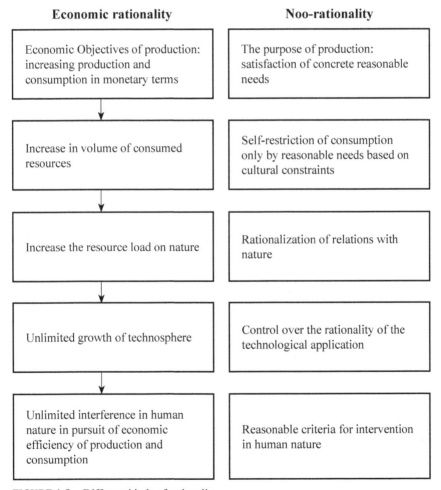

FIGURE 1.2 Different kinds of rationality.

to new production and noonomy, the focus is on satisfying specific and reasonable needs, and criteria of rationality are replacing those based on monetary gain. The needs for knowledge, trust, social recognition, and self-fulfilment, rather than for the absorption of the goods of life (especially material goods) become predominant. The amount of goods consumed is no longer the main objective of human activity because this need has already been fulfilled within a reasonable limit. As G.B. Kleiner writes, "Economic interest as the main criterion for decision-making is giving way to intellectual interest—the demand for new elements of knowledge space and intellectual space."[92]

These targets determine the design of a noo-production regulatory mechanism that does not focus on "noo-GDP" or profit but on indicators that show what we want to achieve. Accordingly, the in-flows that are adequate to the task are formed—information, management, material, and other factors that enable this to be achieved. This is how it should be planned and programmed—e.g., how many of these same flows and controlling influences, and where and when and in what periods to achieve the desired result.

Thus, noonomy does not focus on pursuing profit or other income, which is achieved by a chaotic game of market forces, but rather on the rational desire to *satisfy specific needs, judged as reasonable.* The sensible needs saturation level acts as a specific production target. It implies a program of action that transcends the chaos of the market and gives the production a more structured and orderly character. This approach precludes neither the elements of chance nor the freedom of individual choice not restricted by prescriptions from above. The developed production program must have flexibility and adaptability to changing conditions and random perturbations. This program is to to be adjusted if it does not work, since there are much more factors to be considered in the plan than we can analyze at our current level of knowledge

Let's take a purely conventional example: we have two glasses on the table, and we have planned that we will drink water from them. Then they tell us: next time, a year from now, we will give you another 100% of what we've achieved and place two more glasses on the table for you. Will we need two more glasses? No. But after all, they made them, poured water—and the GDP doubled!

---

[92] Bodrunov, S. D.; On the Issue of Noonomy; *Economic Revival of Russia* 2019, No. 1(59), 5.

This is a tentative example of the absurdity which, in its destructive power, could bring an entire civilization to catastrophe if we were to pursue such a course. This approach, which fosters simulative things, is now often supported by business structures and various government programs. And it used to be supported both in the Soviet planned system and in the non-Soviet one. Each system promoted in its own way a simulative direction of development, a "growth" economic development without "cutting out" the illusory, false component in the structure of needs, as well as without a deep reflection of the plan's goals (often planned simply "from what has been achieved").

That's why we can formulate a principle for the economy (still the economy) of the coming future: *no linear economic growth but economic development.* In this regard, growth is effectively a fiction.

In this context, many of the quantitative indicators that are now used to measure living standards—" to test the harmony by math"—should be relegated to the dustbin of history and replaced by other criteria—a new, different frame of reference—within which to assess the qualitative development of society. However, there is a clear need for a more rational planning mechanism for the solution. (It may be different, as we do not insist on specific methods.) The main, basic thing is the satisfaction of the real needs of people, and recognizing and assessing emerging nonsimulative needs.

But if the market generates a huge number of fictitious needs, what should we do about it? To forbid it is ridiculous and, moreover, impossible. And if not, the economy will recycle more and more of the Earth's nonrenewable resources into fictitious goods.

So, what to do in these circumstances? Apparently, there needs to be a system of thoughtful action and incentives, not just economic ones—if we are moving to a new mode, the economy as we see it now is no longer working. A whole "new normality" emerges. The important thing here is not so much to look at traditional economic indicators but to "calculate" which reasonable individual and societal needs are most effective (least costly, least conflictual, fastest pace, etc.) in moving us toward NIS.2 and beyond.

Only when we can satisfy these needs can we say that "happiness has increased," bring about a better quality of life, not GDP. This task is much less trivial than simple growth planning, which is a concern for economic authorities, and not only in Russia. However, a person who is receptive to the importance of this task, if it becomes a real and perceived need, is in our view already able to begin to address it at today's level of science and technology.

Hardly anyone would argue that human happiness is not about inflating the GDP, or profit, or increasing monetary savings. It is both funny and sad when one says in all seriousness that "happiness is not money, but the quality of life" and when one's reluctance to reach for such results is labelled as "limited rationality." Human rationality is not about choosing solely economic "achievements." As people have wisely observed, "a cottage is a castle for those in love."

Humans are smarter and more rational than "growth ideologues" and numerical volumetric indicators. Because it's not just a need for a phone, a glass, or something else. We care about the quality of the glass, the taste of the water, the "quality" of our own life. And maybe we do not need two glasses, we need just one, but a "nice" one— pretty and convenient with the clean water.

That little nuance—"nice"—is crucial. When we talk about irrational behavior, we choose between two glasses, and we are being forced to choose between two glasses. Better yet, let's smash the old one and throw it away, and here's three new ones "for the price of two." We can take two—it will be more as it's growth—but we choose one because we like it better. The word "like" is not a transcendent or illusory concept; there are parameters by which one assesses size, for example (I take a cup where my finger can fit in the handle), or the elementality of an object which we define as beautiful.

It is actually a different rationality and a different knowledge. In fact, our minds, our rationality, are much broader and richer than the economic limits that today's economic paradigm is trying to push us into.

In this regard, we note that even in today's developed market economy, imbued with narrow economic rationality, a significant portion of goods is distributed for free. There is an important trend here: the further you go, the more you do, the quicker society moves into a new state, the next industrial phase, reducing the value of the product/service.

That is why it is time to abandon the economic growth paradigm and use "growth" parameters as auxiliary ones. It is time to "turn on" the public consciousness while forming an economic model with new ideas about the development of civilization, the economy, and society. The economy and society are inextricably linked. In Soviet times, they used to say "socioeconomic development," but they should have said "economic and social development." And what is development? This is a phasing out of everything that creates a simulative economy today. It is a transition—first and foremost in the "economic mindset."

Therefore, it is inadequate and unscientific to measure the development of a society by the narrow equivalent of GDP and other numerical macroeconomic indicators.

We ought to find other parameters for planning and set planning goals accordingly. And they should be sought through meeting the real needs of people. We should not evaluate by purely physical methods but by qualitative measurements—gauging people's interests, surveys, and eventually indirect research methods. New technologies (big data) provide the tools for such analysis. It's time to move from the arithmetic of trivial addition to metanalysis, even though it's more complicated.

At one time, Club of Rome scientists formulated the thesis that economic growth must be limited to avoid ecological catastrophe. Of course, that is not what they had in mind when they proposed reducing consumption and thereby reducing pressures on the bio-geo-sphere. We can agree that restricting consumption can reduce pressures on nature to some extent (though not necessarily!). Still, the fundamental difference in the theory of noonomy is that consumption of simulacrums needs to be reduced while real needs are increasingly met.

An economy that is focused on numbers, creating more and more things and more capacity, products, and things, without taking into account the real, true need for them, is leading us down a blind alley. People need a different economy, or rather a different "-nomy" a "noneconomic" economy that will match what they really need.

Recall the famous slogan of the alter-globalist movement: "People not Profit."[93] In the twenty-first century it has become almost the main slogan of world social forums.[94] In the context of a new vision, the perception of this logframe is quite positive. Not because these people are selfless or "revolutionaries." The right tone here is not pro-revolution but anti-revolution, progressive, evolutionary, solidarity, orderly, intelligent development. And from the point of view of noonomy's theoretical platform, this slogan is obvious: money is an intermediary. We need to be clear about this: the mediator is destined to leave and the human will take first place. That's why it's "people, not money."

---

[93] Kleiner, G. B.; Intellectual Economy of the New Century: Post-Knowledge *Economy; Economic Revival of Russia* 2020, No. 1(63), 41.

[94] Simic, S. Need, Not Greed, *The Guardian,* January 25, 2007. https://www.theguardian.com/commentisfree/2007/jan/25/post997 (accessed June 9, 2022).

1.4.3 NONECONOMIC WAY OF REGULATING ECONOMIC ACTIVITY

The formation of a noneconomic mode of economic activity will occur in a movement from the modern economic order to NIS.2 and through it to noonomy.

Two stages in the historical process of movement can be distinguished. *The first stage* is the development of *trust technologies*, enabling cooperation without intermediaries, reducing the importance of property, and the socialization of society—this concerns the economic relations between people in the form of needs satisfaction. On this basis, there is a "contraction" of the economic forms of people's activity, the economic institutions that support the link between production and consumption.

In the second stage, labor effort itself disappears as a mediating link between a person's satisfaction with its needs (Fig. 1.3). The Old Testament saying, "In the sweat of your face you will eat your bread," is a thing of the past. This will fundamentally change the nature of the human activity, and the way needs are met—they will become noneconomic.

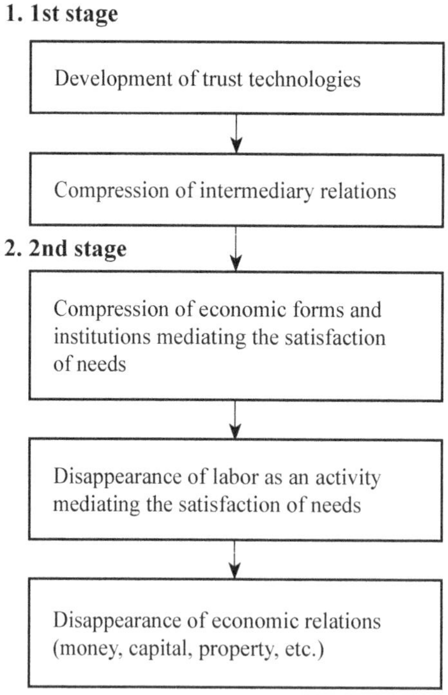

1. 1st stage

Development of trust technologies
↓
Compression of intermediary relations

2. 2nd stage

Compression of economic forms and institutions mediating the satisfaction of needs
↓
Disappearance of labor as an activity mediating the satisfaction of needs
↓
Disappearance of economic relations (money, capital, property, etc.)

FIGURE 1.3 Two steps toward noonomy.

So, in the first stage, we are still in industrial relations and economics, but technology is already emerging to minimize the temporal world of economic relations. The swollen field of mediation, transactional support, etc., will be compressed through trust technologies plus accelerating general technological progress.

And at the second stage, there is no need for an intermediary to meet the needs at all. Roughly speaking, it is not the baker or the store clerk who will meet our needs for buns, but the "bakery" itself. This applies to a host of other professions as well. Human interaction will remain only in the creative process, discovering new knowledge and transferring it to the technosphere and its implantation in new technologies.

But even before the formation of noo-production, *the creative activity that implements knowledge in new technologies, in fact, changes the mode of appropriation.*

The vital difference between the appropriation of knowledge (which, let us remember, becomes a major production resource as early as NIS.2) and a material product is that knowledge, once acquired, cannot be "retrieved" again. It's quite easy with a material object: take it and give it back. And knowledge cannot be "irretrievably" returned.

But the extension of the scope of knowledge also affects the appropriation of material, and not just intellectual, products. With new knowledge and new technologies, the easier, cheaper and simpler it becomes to obtain material goods, the less need there will be for intellectual private property. And in general, the need for the property as an institution will decrease. Not in knowledge, but specifically in ownership. As S. D. Bodrunov states, "What will happen to the "knowledge" part in the future? […] No matter how much we restrict the use of scientific research results by artificial rules, sooner or later, they manifest themselves in the social product, in the social organization, forming a new state of society. We must understand once this fight will stop. But today we are in the first stage of a long transition.

This is the origin of a profound realization, on the one hand of the value of knowledge as a future essential resource. On the other hand, social relations based on the private way of appropriating the result of social production and competition for the resources required for it prevail today."[95]

---

[95] *See,* for example, World Social Forum 2016. http://www.globaljustice. org.uk/events/world-social-forum-2016; A Great Movement Is Born: Global Justice Movement Finds Fertile Ground at the Asia Social Forum, Focus on the Global South. https://focusweb.org/node/144.

This is why they generate those ways of "protecting" intellectual property that "extend" in time the existing social relations regarding knowledge, extending to the field of knowledge the relations that have arisen in the "material" sphere. This stage will, of course, be overcome with NIS.2.

Already at the NIS.2 stage, trends are emerging toward a change, indeed a decline, in the economic forms of human activity, which is evident in acquiring new knowledge. But what will come in place of these economic forms? After all, the sphere of production (albeit without direct human involvement) will not remain as creative, "knowledge-producing" and "culturally productive" human activity, without influence from social relations at all?

This immediately raises a lot of questions. How do people organize their influence on humanless production? How will it be decided where they will send it? What in it needs to be controlled and adjusted? After all, this sphere will exist outside of human relations, but not separately from people, and it will continue to depend on the reproduction of human life.

And here the development of humankind faces a dilemma:

"Either society fails to harness the potential of the technological revolution to improve itself, or it is carried away by false goals and values, exacerbating the negative trends of modern civilization to the point of humans losing their own identity"[96]—and that would mean that we would never enter a new society, never move on to a noo-civilization. Either humanity manages to realize a new approach to the reformatting of current civilizational attitudes or there is no hope for this new society.

But separated from people, from society, it remains subordinated to society. "It is the sphere of setting goals, formulating goals and objectives and controlling the permissible means of their realization in the technosphere that will remain the domain of human society. Autonomous techno-essences, technetical beings functioning in the realm of noo-production and capable of self-development, will depend on human society to determine the limitations of their self-development, blocking directions that are not beneficial to society and orienting the functioning and development of no-production in directions necessary to an individual for its own development"[97] (Fig. 1.4).

---

[96] Bodrunov, S. D.; Coming and Thinking; *Economic Revival of Russia* 2016, No. 4 (50), 15–16.

[97] Bodrunov, S. D.; Noonomy: Ontological Theses; *Economic Revival of Russia* 2019, No. 4(62), 14.

Global Development Trends

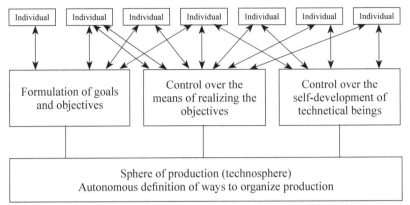

FIGURE 1.4 Relationships of people in the process of regulating the new production.

We are talking about shifts incommensurably deeper than taking environmental constraints into account in economic decision-making. It is the beginning of a qualitative change in the content of production, needs, values and motivation of human behavior and of course socioeconomic relations and institutions. The basis for this, let us emphasize once again, is provided by qualitatively new technologies that turn the semi-utopian modalities of the twentieth century into the practically achievable tasks of the present.

It is not enough to idealize economics or the emerging noo-society, the corresponding production, and the new economic relations, which are not strictly economic in the strict sense of the word. Noo-society does not appear artificially but is an inevitable product of the development of human society at a certain stage. But it *does not in itself guarantee* "the reign of goodness." Therefore, its presence immediately raises the question: *What are the imperatives of the reason that will prevail in it?*

"Hence, the questions-challenges that we have to answer. Hence, the question of the *social form of noo-production*. What imperatives will govern both the production of the material and spiritual conditions of human life, and the social relations that regulate this production? What determines the selection of these imperatives? The state of the noo-sphere will depend on it to a decisive degree."[98]

---

[98] *Ibid.*, 15.

[99] Bodrunov, S. D. From ZOO to NOO: Man, Society and Production in the New Technological Revolution; *Voprosy Filosofii* 2018, No. 7, 114.

As a first approximation, we have already answered this question by stressing the need to bring to the forefront the cultural imperatives of shaping needs and regulating economic activity to meet them. We have designated the social form of such regulation by the term "noonomy," and now we can give its detailed definition.

> *Noonomy* is a noneconomic social form of human economic activity, aimed at meeting noo-needs (mainly the needs in human personality development) based on noo-production development, i.e., such production, which is realized by humans' withdrawal from direct labor activity ("humanless production") and technosphere management as an external to the sphere of achievable human cognitive potential.

Managing social relations will be different, and management itself will be consensual, different in meaning than the present. Even if this system of government could still be called a "state," it would be a qualitatively different state. The main difference between the state of economic society and what comes next is that the state, the current state, first of all, regulates economic relations, and to a certain extent, all other relations. Economic relations will disappear along with the economy, but others will remain, and the regulator is indispensable. There must be institutions of social regulation; the anarchic idea of complete self-regulation is unacceptable. A system of relationships is needed to recognize, compare, and contrast the interests of others. The decision-making framework we mentioned earlier, a cultural framework that evolves with the development of society and the individual, should be developed. So, it is necessary to estimate the pace, the path of development, etc.

In this regard, ways of reaching consensus and consensual management of society should be developed because society is about different interests. There are individual interests, there are public and common interests. And as development goes on, there will be a growing need for such a mode of regulation—based not on economic but on cultural criteria, on the power of human reason, on noo-criteria. Of course, social ties remain as they bind humanity together into a society. But will they be social relations, that is, relations between people as elements of the social structure, as representatives of social classes, social and professional groups, etc.? This type of social relations can also be assumed to have disappeared. There is no basis in noonomy for splitting people into classes, professions (along with the disappearance of the professions themselves) and a general split according to social status.

Moreover, the very sustenance of material conditions of existence is no longer a direct matter of human hands. People will influence this sphere but only by the power of their mind and knowledge.

1.4.4 A STRATEGY FOCUSED ON MOVING TOWARD NOONOMY

Defining the trend of society and social production as a movement toward a new society and noonomy provides the most general definition of the vector of social development strategy. Not only does it clarify the milestones society is moving toward, but it also illustrates how choices are made at the crossroads of civilization as well as how to overcome the risks that uncontrolled and chaotic technological progress and increasing production can lead to.

Noonomy is not a strategic objective in itself. The goal is the development priorities that are embedded in it. It is, above all, a question of the development of the human personality, the formation of a "cultural human"—both as the main objective of production and as the main factor in its progress. The social conditions of production necessary to realize this goal are fulfilled when humans finally withdraw from the nonmaterial production process and the relationships between people in the production process cease to exist. At the same time, the economic criteria of economic activity disappear.

A sphere of "humanless" production is forming, but it is relatively humanless since it remains under the control of people. A system of relations between human society and the technosphere is being developed, where a human being acts as a controlling and guiding external force concerning the autonomous technosphere, implementing the technological application of scientific knowledge in a desired direction.

In defining the strategy for moving toward noonomy, it should be noted that this movement has at least two stages. NIS.2 is first achieved, and then its development reaches an intermediate point where there is not yet a disappearance but a contraction of the economic forms and institutions that mediate the satisfaction of human needs. Thanks to the development of trust technologies, the scope for economic intermediation is shrinking.

It is only later that humans are finally squeezed out of material production, that there is a cessation of indirect labor activity in this sphere, and with it a move away from economic rationality and toward noonomy. The very transition to the frontier of noonomy provides the conditions under which the human personality becomes the main priority and factor of development.

The achievement of this result can be considered as a general definition of the mission of Russia's socioeconomic development, which sets the targets of the strategy. Detailed elaboration of these targets requires examining how moving toward NIS.2 and from it toward noonomy affects the goals and priorities of societal development.

CHAPTER 2

Trends in Socioeconomic Developmental Goals and Priorities

2.1 CHANGE IN THE NATURE OF NEEDS IN THE MOVEMENT TOWARD NOO-PRODUCTION

2.1.1 *EVOLUTION OF WORK AND NEEDS*

The contradictions in the formation and satisfaction of people's needs and the ways to resolve them evolved alongside the emergence of the industrialized mode of production. The industrial production method is based on the mass production of standardized products. This opportunity, in turn, generates a demand for mass consumption. But mass production and mass consumption did not "meet" immediately. It took a series of sharp social conflicts during the nineteenth century and the first half of the twentieth century for mass industrial production to translate into mass consumption, at least in the most developed countries.

The convergence of mass production and mass consumption has entailed an expansion of needs, along with an increase in the capacity to meet them. The technological application of knowledge has made it possible to create new goods and services and increase their output. At the same time, the specific weight of material resources in production decreased, and the specific weight of the knowledge contained in them increased (growth of the *knowledge content* of the product). Were it not for this trend, mass production driven by mass consumption would have collided with absolute resource limits long ago (although this threat has not been removed from the agenda).

The development of science and technology has created a new trend in recent years: products that satisfy several needs simultaneously. In this manner, if growth and consumption are *slowed down* or *reduced*, it is still possible for needs to be *met*. This indicator is also inextricably linked to changes in the nature and structure of needs. These changes are usually attributed to the Maslow pyramid effect, where saturation of lower-level needs allows a shift toward meeting higher-level needs. However, the fundamental reasons for changes in the structure of needs lie *in production and not in consumption*.

The growth of the knowledge intensity of production also implies the same growth in human labor. The exclusion of the human being from the direct production process, the concentration of its functions on control and goal setting, shifts human activity toward a predominantly creative function, linked to the discovery and technological development of new knowledge. For the person in this role, personal developmental needs are paramount—a prerequisite for developing creative abilities.

This change in the content and structure of needs is the most important prerequisite for their saturation. When motivation for personal development becomes paramount, the urge to quantitatively increase consumption of material goods (if it is already secure at a level sufficient to sustain life) diminishes. This change in needs is, in turn, a precondition for and stimulus for creativity in production.

2.1.2 *MEETING NEEDS: SENSIBLE OR SIMULATIVE?*

What human needs will be served by an ever-expanding technological capability to meet ever-increasing demands? And how will these needs be shaped?

Capital always pursues the expansion of mass production and mass distribution. This drive, on the one hand, has led to the constant development of production, improved technology, progress in the forces of production—and, at the same time, an expansion and increase in the diversity of human needs. From the perspective of economic rationality, the kind of needs and means to satisfy them are less important than their ability to attract the solvent demand of the consumer. This is precisely why an industry for the formation and satisfaction of imposed needs has developed along with the progress of production and consumption. But the modern market doesn't just play on human weaknesses to expand production and sales. It creates false, illusory, *simulated needs*, as we have noted, and the means that can simulate their fulfillment.

Simulative needs are illusory, fake needs that are only sympathetically fulfilled and imposed by the market system purely in pursuit of sales expansion.

Thus, "the market economy is becoming more and more a space of production, not so much of real use values that satisfy real needs, but rather a world of creating *simulacrums* that satisfy simulated needs, artificially created by marketing, PR and other technologies that are so widespread with the increasing use of information technology."[1]

Simulacrum goods act as signs of satisfaction of simulacrum needs or means of imaginary satisfaction of imaginary needs.

The nature and role of simulative goods and simulacrums, mere signs of satisfying imaginary needs, have been explored in detail by Jacques Baudrillard[2] from a socio-philosophical perspective.

But the simulacrum is not just a social phenomenon. The mass production of simulacrum has led to the emergence and formation of a vast market of simulacrums, which has become a significant socioeconomic phenomenon.[3]

The law of elevation, expansion, and increase of needs also works in the simulative sphere—as the law of increasing false, fake needs. Because after the achievement of an opportunity to satisfy a need, the following thought arises: What new need can arise? This is due to the nature of knowledge: each "quantum" of extracted knowledge does not only answer the utilitarian question that asks for this "quantum," but also generates new, "complementary" knowledge. This enables new needs to form and the old need to "sprout" into a new one.

But why do these "false" needs arise? Because humans as biological beings, once they understand the world around them and themselves in it, tend to think about reserves and about at least one step ahead in the future. They try to calculate and estimate, based on the accumulated knowledge about themselves and their potential

---

[1] Bodrunov, S. D.; From ZOO to NOO: Man, Society and Production in the New Technological Revolution; Voprosy Filosofii 2018, No. 7, 113.

[2] *See:* Baudrillard, J. *For a Critique of The Political Economy of The Sign,* Editions Gallimard, 1972. Russian translation: Baudrillard, J. Toward a Critique of the Political Economy of the Sign, Biblion-Russian Book, 2003.

[3] For an analysis of the nature of simulacrum goods and the market for such goods, see: Buzgalin, A. V.; Kolganov, A.; Simulacrum Market: a View through the Prism of Classical Political Economy; Alternatives 2012, *No. 2,* 65–91.

needs, what they will need. And when there is an opportunity to create reserves, they do so.

Hence—trivial as it may seem—comes all the ideology of hoarding for something, the ideology of getting extra space that you don't need now and may not need later either. Gradually, the right need for potentially useful things, i.e., the need for natural accumulation, starts to cross the line where one does not know exactly how much one needs but realizes at some point that even this may not be enough.

But this natural need, in a world where the conditions of existence are precarious and uncertain, does not have strict boundaries. Any stock, any amount of what has been acquired, seems insufficient; in such conditions, at least some confidence in the future begins to be measured by the size of the "mountain" of accumulated goods. This aspiration is also socially reinforced as the accumulation of wealth becomes a symbol of success, of one's social status, and the pursuit of this status is identified with an increase in the amount of goods received (although perhaps not actually consumed).

Thus, simulative needs grow in tandem with the satisfaction of normal needs. But there is already a distinction: a simulated need can be fulfilled even though it is illusory in nature, i.e., a person does not really need that many useful things, not now, not in the foreseeable future, nor at any other time. And it can be useless too. Nevertheless, it is possible to meet such a false, "imposed" need. There can, of course, also be needs that are pure simulations of rational needs that *cannot be met* at this stage in principle, but which can be thought of. These can be called phantasms, whereas the first type of simulated need is a redundancy (Fig. 2.1).

It is important to remember that in certain cases, simulative needs may become nonsimulative and vice versa. At the same time, a real need for one may be simulated for another. For example, the need for a tailored dress or facial care products for a peasant woman in past centuries was more of a simulative need, whereas today it has become the norm; while the logarithmic scale that every engineer needed at the time is now probably only needed by a collector of old measuring instruments.

We are actively moving toward the increasing satisfaction of more and more irrational needs. The entire current economic paradigm is set up to do just that.

Trends in Socioeconomic Developmental Goals and Priorities

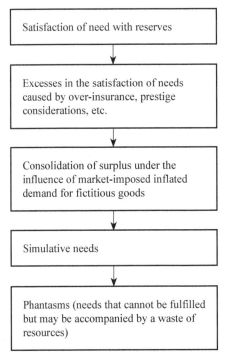

FIGURE 2.1 Formation of simulative needs.

As S. D. Bodrunov states, "It has to be considered that the modern market economy resorts to extraordinary inflation of simulative needs in pursuit of sales volumes. It is no coincidence that the production and consumption of simulacrum have spread so widely in recent decades. The underlying reasons for this are the shifts in the structure of social production that took place at the turn of the 1970s and 1980s when myths about the post-industrial economy swept the world. They did not emerge out of thin air: the *unrestrained growth of the service sector on the one hand, deindustrialization on the other, the virtualization of everything and everything that fuels it, are the material basis for the expansion of* simulative production *and the spread of simulative needs.*"[4]

Simulative consumption is not simply a matter of personal choice. Inflating simulative needs in the real economy means spending more real resources on things that give only the illusion of usefulness. And the

---

[4] Bodrunov, S. D; From ZOO to NOO: Man, Society and Production in the New Technological Revolution; Voprosy Filosofii 2018, No. 7, 111.

growth of simulacrum production is one of the significant components of the increasing resource burden on the environment.

2.1.3 THE NATURE OF NEEDS IN NOO-PRODUCTION IS THE NATURE OF PRODUCTION AND THE NATURE OF NEEDS

The universal nature of human beings, the challenges of the technological revolution, the development of new needs and new ways of meeting them also linked to the new technologies—where are these trends leading humankind? And what direction are they taking the current economy in?

As has been shown above, the possibility of increasing the satisfaction of needs at decreasing costs creates both the possibility of reducing the resource burden on the biosphere and the temptation to overabundance. What will be the choice of humanity? What will it be defined by?

What will happen?: Will the "human of profit" be replaced by a "human with other motives" (self-development, quality of communication, and social recognition)? Or will people drown themselves in a sea of increasingly sophisticated and illusory pleasures? Or will people hide from life in a virtual digital space? After all, virtual space can both empower and constrict communication, serving to isolate the individual think of the Japanese "hikimori," or recluses, who have not left their computers for years, who not only refuse to talk normally, but even to go about their normal daily routine of eating, getting dressed on time, and maintaining physical fitness.

If we get past this fork in the road and find ourselves in "noo-production" it would be more *the "production of people themselves" than the production of the material conditions of their existence*. "The structure of human needs will also change accordingly. Self-development needs, spiritual needs, need for communication, social recognition will be predominant. These needs will determine the technology used, the products manufactured, and the organization of production to meet material human needs. These shifts in the structure of needs will be determined by the progress of human culture"[5] as specific knowledge.

"In summary, *social production in a noosphere society*, as far as we can tell from the analysis of objective processes that have already begun to develop in recent times, is formed as a system that includes:

---

[5] *Ibid.*, 115.

- the priority development of knowledge-intensive, "intelligent" production (we can call this *noo-production*);
- the resulting integration of production, science, and education within the framework of unified reproductive contours, leading to the formation of a new type of reproduction called *noo-reproduction*, which provides the priority formation of conditions for the development of the noosphere;
- the gradual diminution of utilitarian and simulative needs and the rise of a new class of needs—the needs of the "rational human," or *noo-needs;*
- the development of new, appropriate values and motivations for the main actors of material and spiritual production, which have the characteristics of economic production;
- in the transition period, economic relations and institutions transforming toward socialization and humanization, through active development of *noo-oriented* economic programming, active industrial policy aimed at the priority development of smart production, and strengthened public–private partnerships focused on these tasks;
- and, finally, the elevation of culture as a key enabler of *noo-development.*"[6]

Noo-production is knowledge-intensive production, which minimizes direct human involvement and focuses on meeting the *noo-needs* of the individual, creating the conditions for human advancement in the realm of knowledge and culture.

To summarize, the main goals of noo-production can be defined as the *growth of the individual.*

This growth of the individual is the explicit aim of production in society, to develop human qualities and enlarge the human cultural space regulated by culturally produced values.

This will occur through the growth of spiritual needs in all areas of human culture. An important part of the need for personal growth will be the need for conscious self-limitation of simulative needs (which, along with the use of new technological possibilities, will make a significant contribution to the implementation of the resource-efficient development path).

---

[6] *Ibid.,* 114–115.

This self-limitation is not an external imperative, although, in the transition to a new society, external moral imperatives, explanation, persuasion and, finally, cultivating the habit of reasonable self-limitation will also play a role in limiting simulative needs. This will certainly be enhanced by accelerating technological progress, which, through noo-industrial production in the NIS.2, increasingly devalues the material, tangible product, making the satisfaction of vital and other nonsimulative human needs less and less important to the individual for this process. It will become more and more valuable to satisfy the growing spiritual needs of people.

Of course, what is most effective is the *inner* self-limitation that grows out of the determination of the needs structure by the new nature and content of the human activity and the social relations in which it will be carried out. Even today, people engaged in, say, deciphering the human genome or developing the technology to send an expedition to Mars are not likely to be concerned with buying villas on the Côte d'Azur or huge oceanic yachts as a priority, regardless of their income level. For people passionate about this kind of work, such needs are irrelevant because meeting them would only hinder rather than help them achieve the goals they set for themselves.

The quality of the spiritual, cultural component of human development should define all other directions of human development and subordinate them to the best norms of human culture.

A noo-civilization must be a sustainable one. The system should be resilient and work to increase its resilience, to maintain itself as a system, and not to break down.

Preservation of the self as a human being, i.e., preserving a system in which the developed human being is the basic element of sustainability, is a clear and understandable goal. An individual of the noo-society is an element of this society as a system, which allows this system and civilization to be sustained, ensuring the sustainability of its development, which is, in general, the basic value of existence. If it is not a "new" person, living according to the criteria of a new society, but an "old" one, adequate to the "old" system, the system will become different and technotronic (with the negative consequences described above).

An individual's development as a personality can only be preserved in an environment where they are aware of what they can and cannot do. What is needed are basic institutions that do not support what is happening today within the system of global capitalism but are designed to provide a new way of development.

And here, strange as it may seem, the development of technology is required. The accent, however, must be placed on cognitive and social technologies: nano-, bio-, and information technologies, this is the cutting edge for the 1990s, i.e., to some extent "worked out" material. Shortly, we must move from information technology to cognitive technology, and then to a combination of technological and social knowledge; otherwise, if we don't develop the ability to know ourselves and the world better, to absorb the vast amount of knowledge being generated, we won't be able to "implement" this new way of development.

It is only in this scenario that we can be assured of a future.

2.1.4 THE CULTURAL HUMAN AS A PRODUCT OF NEW PRODUCTION AND INDIVIDUAL GROWTH

When (with a restriction/decrease in need) productivity increases, the length and importance of working time decrease, and the value of free time increases. NIS.2 can already provide a significant increase in free time, but it will not provide a similar "increase in happiness" immediately—you still need to learn how to channel your free time into self-development (growing spiritual needs, culture, etc.)

Hannah Arendt's scepticism on this issue is understandable: she doubted that more free time would ensure human development, since, according to her, people tend to use this time only for mindless consumption. As Arendt writes:

Animal laborans never spend their excess time on anything but consumption, and the more time they are given, the more insatiable and dangerous their desire and appetite will become. Of course, lust has become more sophisticated, so that consumption is no longer limited to what is essential, but rather to what is superfluous; but this does not change the nature of the new society, and worse, it contains the grave threat that in the end all objects of the world, the so-called cultural objects as well as objects of consumption, will be devoured and destroyed.[7]

Yet with the kind of social order we are living in, so-called capitalism, this is exactly the case, because capitalism only leaves free time for people

---

[7] *See:* Arendt, H. *Vita Activa, or On the Active Life,* Aletheia, 2000, 171.

to consume what they produce during working hours, then earn again and consume again, etc., pushing them equally hard to consume as to produce for the sake of that consumption.

Society can find a way out of this vicious circle, not through an ideology of asceticism, forced rationing, a reduction of consumption, or propaganda for higher ideals, but by reducing the time needed for work (preconditions which are already present in modern industrialized production methods) and simultaneously developing creative activities in one's leisure time.

It is the opportunity for free activity that sets the stage for the voluntary and intelligent choice of an appropriate structure of needs and lifestyles. To quote Nobel laureate economist Amartya Sen, sociocultural factors of primary education, basic health care, and employment are most important because of the role they can play "in enabling people to engage with society decisively and freely. Such problems require a broader information base, focusing on people's ability to live according to their own reasoned choices."[8]

But we should not think that Hannah Arendt's doubts came out of nowhere—*the transition from free time as a time of consumerism to free time as a space for human cultural development is not a quick and easy matter.* This is a problem of colossal importance, capable of generating difficulties of very grave scale and depth. Only its solution finally brings us into the era of noonomy.

The individual in NIS.2 is in many ways able to emerge not as a mindless consumer but as a creative individual, since the creative use of free time depends largely on the material prerequisites for creative activity access to means of self-education, physical improvement, scientific and artistic creation, etc.

Of course, a change in the ratio of working time to free time is also a prerequisite. And the transition to the next stage of noo-production poses, more than ever, the immense and profound challenges of acquiring new knowledge to make a leap forward in technological advancement and to realize the direction and limits of one's own development. It is the need for these tasks to be solved, as well as the practical involvement of the individual in the technological (and sociopractical) application of science, that will determine the face of free time in the new societal phase.

While Arendt derived her conclusions from the observation of real social contradictions in her society, she did not consider an important law:

---

[8] Sen, A. Development as Freedom, Novoe Izdatelstvo, 2004, 81.

a change like human activity, aimed primarily at acquiring new knowledge gradually and not immediately, will also change its needs, their structure and qualitative content, therefore filling up *free time*.

It is communication—information—and the knowledge contained within, that will become more valuable than the most important material values before. And we are no longer that far away from that prospect becoming widely understood.

This new society will undoubtedly bring about fundamental shifts in the way most people live. Former occupations and professions will lose value, and the transition will be harrowing. The agrarian revolution in Britain in the sixteenth and seventeenth centuries produced many beggars and vagrants who were brutally repressed, while the industrial revolution of the eighteenth and nineteenth centuries involved the mass ruin of small craftsmen and the suffering of the "industrial reserve army." But in both eras, there was no social catastrophe. The landless peasants were either converted into wage agricultural laborers or absorbed by the growing manufacturing industry. Bankrupt craftsmen joined the ranks of the rapidly expanding factory proletariat.

Thus, the *coming technological revolution will make entire professions redundant, create new jobs and, subsequently, types of occupations*. New technologies will give rise to new needs and meeting those needs will require new technologies. At the NIS.2 stage, new jobs will emerge to replace those "eliminated" by automation and productivity growth. In addition, the inevitable rise of the knowledge economy (in a transitional phase) and the increasing need for new knowledge may absorb many workers.

However, with the change in the technological basis of production and the transition to noo-manufacturing, the very concepts of "profession" and "workplace" change their meaning dramatically, if they don't disappear altogether. The profession as a way of earning money through certain skills is likely to become a thing of the past. These functions will be performed by technetical beings, while "an individual will aim to develop an approximation to absolute knowledge and will be increasingly universal. New ways to access knowledge and information—networks and various man-machine systems—will be developed.

Of course, human universality will not consist in everyone knowing everything, but in the possibility and ability to master almost any necessary knowledge. *The main shift will be creating information and communication systems that allow everyone to benefit from the whole ocean of knowledge that humanity has accumulated."*[9] *And a person will penetrate into its greater depths.*

"It goes without saying that this requires the development of one's own abilities, the ability to enter and navigate every field of knowledge. Such universality is within reach if the educational system is restructured accordingly"[10] and the natural capacity of human is enhanced. The main task of the education system "will not be "pumping" the student with knowledge and skills in a particular narrow specialty. The learner must cease to be a passive recipient of ready-made knowledge and learn to "acquire" and apply this knowledge independently. Of course, this skill cannot be acquired without a broad foundation of education, enabling one to orient oneself rapidly in any required field of knowledge.

The transitional step to such a "universally self-learning" person is the achievement of the concepts of "education for all" and "education through life," which are necessary to reach the stage of NIS.2.

The development and availability of new, increasingly sophisticated and universal means of access to knowledge is then of critical importance."[11]

2.1.5 HOW DO CHANGING SOCIETAL NEEDS AFFECT STRATEGIC DEVELOPMENT GOALS?

As we have established above, modern society is characterized by the presence of both rational needs and fake, illusory, and simulative ones. The latter's presence is a problem that is closely linked to the nature of existing economic relations, their criteria of rationality, and the consumption pattern that they shape in humans.

This demand pattern has already created serious risks of inflated resource consumption leading to excessive pressures on the natural environment. Technological advances can significantly increase the satisfaction of needs while reducing the specific consumption of resources. But this opportunity will only reduce the risk of anthropogenic destruction of the natural environment if it is not used to increase the volume of simulation needs. Only by moving toward satisfying reasonable needs, which

---

[9] Bodrunov, S. D.; Convergence of Technologies—A New Basis for Integration of Production, Science and Education; Economic Science of Modern Russia, 2018, No. 1 (80), 16.

[10] *Ibid.*

[11] *Ibid.*

include reasonable self-restraint, can the contradiction between growing needs and increasing pressures on the natural environment be reduced.

But is it possible for society to set as a strategic goal the limitation of human needs? Isn't the goal, on the contrary, to satisfy them as much as possible?

These two approaches are not alternatives if (1) self-limitation concerns simulative needs and (2) the possibilities of satisfying reasonable needs are expanded. Therefore, the strategic goal of societal development becomes the creation of opportunities in which these two conditions are met. This is possible when the rational economic individual who maximizes consumption is replaced by a cultural individual. This is a person with a changed structure of needs—a shift from absorbing more and more material goods to satisfying the need for self-development. The precondition for this shift is not only access to education and a culture of knowledge; this is a necessary but not a sufficient condition. The decisive factor is the change in human activity, from labor dictated by need and economic rationality to creative activity that develops the human personality.

This is one of the most important strategic development goals because it ensures a shift toward sustainable consumption and a potentially limitless opportunity to harness knowledge and apply it to production development.

2.2 QUALITY OF LIFE AS A TARGET FOR SOCIETAL DEVELOPMENT

2.2.1 TRANSITION FROM THE NEEDS OF AN "ECONOMIC INDIVIDUAL" TO NOO-NEEDS

Cognition of the external world and the self involves accepting limitations. By defining oneself as intelligent, one *sets a boundary*, separating oneself from unintelligent beings. At the same time, it is inherent in human beings to move toward, and go beyond the limits they have not yet reached. But it is only when an inner boundary regulates this aspiration that it is productive and creative, not destructive.

A qualitatively new noo-industrial needs satisfaction mechanism will be based on the new nature of the reproductive link between production and consumption. Human needs, as well as the knowledge needed to fulfill them, will be shaped not by direct productive activity (for the human to emerge from it), but by creative self-development. Such needs can be referred to as *noo-needs*.

Noo-needs—needs to be determined by the criteria of human reason and cultural imperatives, based on a rational level of satisfaction of vital needs and the increasing role of higher-order needs.

These needs and this knowledge will constitute the "order," or the "terms of reference," for an autonomously functioning "humanless" sphere of direct material production. By transmitting this order to noo-industry, individuals will end up with the necessary tools to satisfy their needs, without being directly involved in the process of producing these tools or in organizing it. These tasks will be solved by a relatively independently functioning technosphere.

As the content of human activity and the nature of needs are changing, so are the criteria for the rationality of consumption and the structure of needs. Economic criteria of rationality are replaced by reasonableness of needs criteria defined by human culture (Fig. 2.2).

The human orientation toward purely economic criteria of success must, in the long run, disappear not only because the structure of human needs is increasingly based on motives and values that are not amenable to cost measurement and often cannot be defined in terms of cost-benefit ratios at all.

Economic rationality is also increasingly questionable because of its negative consequences. It deforms the structure of human needs, trying to fit them into the Procrustean bed of monetary symbols of success and giving the status of rational only to those and exactly those achievements that entail the growth of cost measures formed by the market.

And is every increment of monetary wealth for good, and should everything that has no monetary value be rejected as irrational? Experience has shown that market prices fail to capture—or can capture with great distortion—a great deal of important factors in human existence and development, such as environmental dynamics. Is it possible to put a monetary value on the loss of biodiversity? Is it possible to measure in money the level of human culture or the value of human communication?

2.2.2 QUALITY OF LIFE: WHAT IS IT?

Thus, *it is not the quantitative increase* in consumption that becomes the goal of production but the improvement of the *quality of human life*.

Trends in Socioeconomic Developmental Goals and Priorities 75

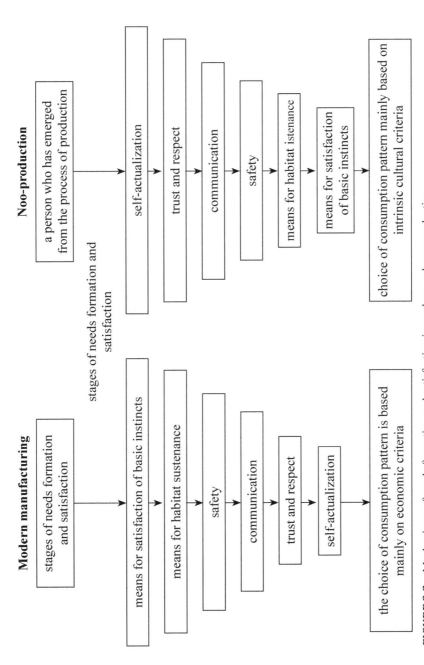

FIGURE 2.2 Mechanisms of needs formation and satisfaction in modern and new production.

Quality of life is a broad and rather abstract category, which varies in content at different historical stages of societal development and its particular manifestations depending on the civilizational features of society. When we look at the quality of human life in the transition from the pre-industrial to the industrial era, we witness a dramatic change, with a more secure supply of varied foodstuffs, the move to industrial housing and the provision of public services, the possibility to make use of the fruits of standardized mass industrial production of clothes and shoes, the provision of basic sanitation and hygiene, and the dramatically increased possibilities of transport and communication.

The transition to an advanced industrial society (including both the new industrial society and what some theorists saw as "post-industrial") has led to greater opportunities for consumption but has also revealed dubious criteria against which its extent has been measured.

The developed industrial society is marked by an orientation toward quality of life, which is connected not only with the growth of consumption of goods that undergo constant technical improvements, but also with a sharp increase in the consumption of durable goods (furniture, household appliances, cars, etc.) and "mass culture" products, while the volume and proportion of consumption of various services is increasing. This consumption pattern is causing the increasing absorption of natural resources, both to produce more and more material goods and to ensure the functioning of a disproportionate service sector.

At the same time, the social structure in the industrial phase of development was characterized by considerable inequality in consumption. It was objectively unavoidable, acting as a harsh but necessary stimulus for production development, multiplying labor efforts and entrepreneurial activity. In the long run, however, this inequality can become a constraint, limiting the unfolding of one's creative potential.

The transition to a noo-consumption model is associated with a change in the content of the category "quality of life" and in the criteria for human development as a person.[12]

This goal does not renounce the diversity and richness of human needs, nor does it imply an abandonment and acceptance of an ascetic ideology. On the contrary, it can only be achieved by developing human universality

---

[12] Kvint, V. L.; Okrepilov, V. V.; Quality of Life and Values in National Development Strategies; *Vestnik of the Russian Academy of Sciences* (formerly: Vestnik of the USSR Academy of Sciences), 2014, *V. 84, No. 5*, 412–424.

in both production and consumption. Only the criteria of human wealth are shifting—moving away from economic rationality to cultural rationality. The concept of *quality of life* embraces the conditions for human development, a favorable living environment, and a high level of culture and human interaction.

This consumption model does not obviate inequality. However, it is not unequal access to any goods (because free access to them is achieved), but unequal abilities realized in developing one's own creative potential and unequal abilities to master the wealth of human culture.

The ideology of marketization has gone so far as to consider even human communication only as a factor in capital growth. But does this mean that we should put a lid on everything that does not add value and declare everything that does? If one accepts this view, one can certainly say where humanity will turn at the fork of civilization. If they are well paid for, any needs —the most fake, the most perverted—will be prioritized. If there is demand, there will be supply.

But humanity cannot be "taken by the hand" and led away from this road. It will have to make a conscious inner choice, shaping new criteria of rationality based on expanding the field of knowledge and the wealth of culture created by humanity. This would be a step toward a noo-social stage of evolution, where the human mind would gain sovereignty over spontaneous, subjugating socioeconomic processes.

2.2.3 *STRATEGIC FOCUS ON QUALITY OF LIFE*

The quality of life of any socioeconomic strategy is essentially its main reference point.[13] The strategy for socioeconomic national development therefore requires, first of all, the development of criteria and characteristics of quality of life that go far beyond the sphere of consumption and cover all the basic aspects of creating conditions for the intellectual and creative development of the individual. For the re-industrialization phase, this goal will be one definition, the movement toward NIS.2 another, and the transition to noo-community a third.

But whatever the definition of quality of life in these stages, it will evolve substantively in the direction of personal growth of the individual rather than an increase in material consumption. While addressing the

---

[13] Kvint, V. L.; Economy in Industry, 2020, No. 3.

challenges of re-industrialization, which does not lead out of the world of economic rationality, will inevitably involve an appropriate model of consumption, the concept of quality of life will at this stage move away from a consumption model oriented toward the "economic man." It is essential already at this stage to develop the conditions for an empowering and creative "cultural individual."

CHAPTER 3

The Basis for Strategizing National Development

3.1 STRATEGIZING AS A METHOD FOR IDENTIFYING DEVELOPMENT INTERESTS, PRIORITIES, AND GOALS

3.1.1 PREREQUISITES AND CONDITIONS FOR STRATEGY FORMATION

As V.L. Kvint writes in *The Concept of Strategizing*, "Strategic thinking, as well as philosophical thinking, derives from many traditions and schools of varying levels of maturity. If the ontological approach to the analysis of facts and Aristotle's concept of the good life predetermines the strategy's focus on improving the quality of life, a related category of the concept of the good life, then another philosophical trend, existentialism, which states free choice as one of the cornerstone issues, predetermines this category as one of the principles and postulates for the strategy and the conditions for such free choice in the strategizing process."[1]

Justifying new strategic perspectives, selecting priorities, and developing scenarios in an environment where the past is only partly extrapolated into the future, the "present" does not exist, and future social processes and economic agents remain largely unknown is a challenge even with a theory that provides a long-term vision. Facts and current technologies, which always refer to the past, are needed for analysis and not necessarily for inclusion in a strategic scenario for the future. Yet, concrete examples and axioms from the past can

---

[1] Kvint, V. L. *The Concept Of Strategizing: a Monograph*, Kemerovo State University, 2020, 24–26.

be used to prove the recurring links of phenomena and processes that predetermine the fundamental principles and laws of strategy.[2]

A strategic approach to national development is primarily about using strategic ideas rather than the immediate use of arms, capital, natural resources, or labor.

Thus, strategy development depends to a large extent on the quality of the strategic ideas underlying it. And the longer the strategic horizon, the more important it is to choose the right underlying concepts and the degree to which they are scientifically sound.[3]

The idea of noonomy pushes the strategic horizon into the distant future. Can we at least give a rough indication when this future is coming? Not yet. But is it possible to build a strategy without reference to a specific time scale? The theory of noonomy, then, applies to strategy to the extent that it allows for identifying goals that can be evaluated in terms of the time required to achieve them.

The theory of noonomy is not a tool for calculating the exact timing of turning points in development and achieving the goals corresponding to these points. However, it makes it possible to determine the logical relationship between events and, therefore, the movement sequence toward these goals.

Can the theory of noonomy help in defining these kinds of goals? Yes. As far as national goals are concerned, these are the necessary intermediate stages of the country's movement toward noonomy: the reindustrialization of Russia on a state-of-the-art technological basis and, eventually, entry into the space of a new industrial society of the second generation (NIS.2). The theory of noonomy is vital when strategizing the achievement of these goals because their content and specificity depend crucially not only on the direction that strategy and noonomy theories take but also because they complement each other in the processes of developing a strategy methodology.

To develop a strategy, the first step is to focus national development projects on the long term, to set up the design of such projects to look for unexpected asymmetric solutions and to ensure that innovations are

---

[2] Kvint, V. L. To the Analysis of Strategy Formation as a Science; *Bulletin of CEMI RAS* (Online) 2018, *No. 1*.

[3] Kvint, V. L. *Strategy for the Global Market: Theory and Practical Applications,* Routledge, Taylor & Francis, 2015.

selected and used to create or strengthen their competitive advantage.[4] The theory of noonomy concentrates on such innovations, making it possible to assess their importance concerning both immediate and very long-term prospects.

From this perspective, Russia's development prospects cannot be reduced to eliminating shortcomings that hinder achievements of advanced scientific and technological breakthroughs and reducing the competitive gap in these areas. Russia's competitive advantages should be accelerated in areas where we have them. Such a push is designed to take us to the frontier, not just the present but also the future, to capture and retain leadership in these areas.

A strategy should establish a clear understanding of global patterns, identify the true values and interests of the target, formulate priorities, assess whether they provide a competitive advantage, set goals and objectives in line with the priorities and determine the most effective ways to beacon its vision of the future before competitors see these strategic perspectives.

The theory of noonomy allows us to look at objectives whose very existence has not yet been recognized by our competitors. This vision of goals derives from an understanding of our interests and values based on the theory of noonomy. These are, first of all, the values of human personality development, based on the totality of the benefits of human culture. It is these factors that the theory of noonomy considers as key to the progress of social development.

Implementing the strategy involves continuous evaluation of past factors and forces, extrapolation of known axioms and patterns, analysis of innovations, new technological solutions, and assessment of their impact on previous future scenarios.

Continuous monitoring of science and technology trends is crucial to the development and implementation of strategies formulated simultaneously based on the theories of noonomy and strategy. Equally important is assessing the impact of these trends on the natural environment, the social fabric, and the individual. This ongoing reflection allows the most relevant empirical inputs to be effectively used, providing a solid foundation for theoretical and then methodological *visions* of the future. However, the long-range horizon of strategic decisions requires the ability to abstract

---

[4] Kvint, V. L. *The Concept Of Strategizing: a Monograph*, Kemerovo State University, 2020, 48.

from current reality and to resort to intuitive identification of the characteristics and conditions that are most relevant to the future.

The strategic development project should clearly present the external and internal environment of the strategic object, which will have developed by the time the new strategy is implemented.[5] The conditions and data obtained are used as starting platforms for the development of the strategy. Without a strategy aimed at long-term sustainable success, the use of new technologies often leads to only temporary victories.

In the context of the theory of noonomy, the search for and the capture of fundamentally new technological niches is of paramount importance. Only the most advanced technologies, the products of new knowledge based primarily on the use of human intelligence, open the door to the future. The most important technologies in terms of noonomy are those that help to displace people from the direct process of production and move them into knowledge-intensive activities, including the research, achievement, and technological application of knowledge.

A strategy leads a company, government, or any strategist from the past to the future, based on anticipation, foresight and strategizing, ensuring that new opportunities for success are responded to and exploited, pointing out potential and poorly known challenges and obstacles to the future, avoiding the latter where possible.

The theory of noonomy acts as a concept that describes the fundamental characteristics of the movement from the past to the future, including our country. This theory points out the major problems and threats we already face and those that are yet to come. Among the first is the technological backwardness of most developed countries, the global problem of exceeding environmental carrying capacity, and the poorly controlled evolution of the technosphere. Among the latter, the threat of irrational interference in human nature and the inability to make decisions that address the already visible threats are beginning to emerge because of the predominance of narrow criteria of economic rationality.

The strategy should be based on the anticipation of potential developments, a vision of the future, and defining the political, economic, technological, environmental, and other conditions of the future in which the strategy will be implemented and leading the object of the strategy to success.

---

[5] Kvint, V. L. *Strategic Management and Economics in a Global Emerging Market*, Business Atlas, 2012, 387–389.

To provide the strategic project with this vision of the future, the theory of noonomy develops a holistic picture of interconnected conditions, the achievement of which helps to guide Russia's national development toward a successful transition to a better future.

3.1.2 NATIONAL STRATEGY: STRUCTURE AND LINKAGES

National development strategies always take place in a global, interconnected world and must therefore be sensitive to the forces and contradictions of globalization. "Globalization as a pattern has very prominent cultural and religious implications of long-term impact that must be understood, appreciated and exploited by strategists working in the global marketplace (GM) and its national and regional subsystems."[6] These implications are clearly visible in the conflict between global and national trends, in the tensions between national interests and transnational capital interests, in conflicts between the trend toward cultural leveling and the protection of national-cultural interests.

Globalization, however, should not be seen as a process that fatally absorbs the capacity of nation states to develop their own strategies. "In fact, there are always alternatives, which by their very nature can determine the limits of nation-state action in the global system."[7]

The theory of noonomy emphasizes the fundamental importance of cultural values as an increasingly important and, in the long term, decisive factor in regulating the entire life of society (including economic life). Two paradoxical cultural dynamics have to be "considered in this regard: the global conversion of cultures and at the same time the preservation and protection of national and local cultural identities and values.

Cultural and religious risk as phenomena, factors, and strategic categories can be fruitfully studied through the joint efforts of strategists, economists, culturologists, and theologians, while ignoring the impact of these phenomena on strategy development and implementation reduces its effectiveness and leads to complex negative economic and social consequences, including those associated with extremism and terrorism.

---

[6] Kvint, V. L. *The Concept of Strategizing*, RAS-HSIU, 2019, 25–26.

[7] Amin, S. *October 1917 Revolution, a Century Later*, Daraja Press, 2017. *See also*: Amin, S. *October Revolution 1917, One Hundred Years Later*, Bodrunov, S.D., ed.; Cultural Revolution, 2018.

The rapid pace of urbanization in emerging economies is further widening the cultural divide."[8]

The success of a strategy stems from the initial anticipation and vision that helps strategists "recognize emerging patterns, trends, and competitive advantages and anticipate their impact and effectiveness before their competitors and adversaries."[9] The future-oriented vision, which is being developed in the theory of noonomy, makes it possible to take long-term development trends and link them to concrete steps in science and technology, economics, management, culture, and many other fields. This gives strategy development the advantage of being able to look far and set goals that go beyond their vision of the future.

A common mistake in national and regional strategy development is to ignore the strategies of domestic and foreign corporations operating within their territories. It is the corporate strategies that should translate the strategies of countries and regions into reality. The importance of developing corporate strategies that flesh out national strategic projects requires that these strategies be coordinated with the overall vision of the national strategy. Thus, corporate strategies should be based on the need to participate in implementing certain goals of the national strategy. Such subordination should not be enforced but rather through the corporations' coordination, economic interest, and initiative.

Developing a new strategy or revising an existing strategy should "begin with the analysis of mature and widely accepted patterns and trends that are directly relevant to the subject and the monitoring of the dynamics of their effects."[10] More importantly, the strategy should anticipate patterns and trends that have not yet emerged and strategize on their cinematics and potential impact accordingly. *"The most innovative and potentially successful strategies are based on an analysis of trends and patterns that are little known or not yet recognized at the outset of a strategy."*[11]

Therefore, it is necessary to take advantage of the opportunities provided by the theory of noonomy by showing the significance of trends that are now regarded as of little importance, but which are destined to play a decisive role as we move into the future. It is crucial to understand

---

[8] Kvint, V. L. *The Concept of Strategizing*, NWIU RANEPA, 2019, 27–28.

[9] *Ibid.*, 26.

[10] Kvint, V. L.; Strategy Development: Monitoring and Forecasting of Internal and External Environment; *Management Consulting* 2015, *No. 7(79)*, 6.

[11] *Ibid.*, 6.

the progressive displacement of people from direct production, the gradual weakening of economic activity criteria, and the increasing importance of cultural criteria. Human beings are the ultimate goal of production, and their development, based on the progress of knowledge and cultural growth, is the most important factor of production. Russia's national priorities must shift in this direction and implementing such a change has a poorly understood but compelling strategic advantage.

The strategy should analyze the development of regional and sectoral economic structures, scientific and/or military capabilities (depending on the object of the strategy); the pace, proportions, and vectors of development to identify potential new opportunities as soon as possible and potential threats. However, a fundamentally different analysis is needed to identify the underlying interests and national priorities that need to be localized in a particular region. It should start with identifying those competitive advantages in the region and/or sector strategy that, if resourced, can contribute to the realization of a priority of national relevance.

This question assumes a shift from a "framework" assessment, set by the fundamental conclusions of the theory of noonomy, to identifying those potentials for Russia's development which can be seen as stepping stones bringing us closer to the achievement of strategic priorities. Selecting which competitive advantages to exploit as a priority also depends on tactical considerations. However, the main selection criterion should be the contribution of these competitive advantages to achieving Russia's strategic development goals.

The strategic plan should be monitored during its implementation to avoid unforeseen obstacles, reduce their negative impact, and respond to obstacles and complications that cannot be avoided. It is also vital to include leaders and managers in the monitoring of strategic planning processes to ensure that implementation time is monitored and to encourage effective strategy implementation.

"Even the most successful strategies reach a stage at which, due to a change in conditions or needs, a transition to a new strategy must begin. If the strategy leads to failure or is foreseen to fail shortly, the implementation of the strategy should be discontinued or adjusted. However, even if strategic success is achieved, the object of strategy must be deliberately led to the destruction of the achieved balanced state to achieve a new level of balance under qualitatively new and more effective conditions."[12] This

---

[12] Kvint, V. L. *The Concept of Strategizing*, RAS-HSIU, 2019, 88.

prevents the object of the strategy from stagnating and moves it toward new priorities and progress.

The envisaged strategy for Russia's development based on the concept of noonomy is distinctive in that it sets the stage for the transition from one strategic stage to the next. Success in realizing the immediate goal of reindustrializing Russia based on the latest technology will immediately form a coherent system of NIS.2. In turn, as society moves in this direction, it will face the challenges identified in the broad outlines of the theory of noonomy. Each of the phases will require different strategies. Thus, the movement toward noonomy *is not one strategic project but a strategic design encompassing a successive series of such projects*, the most distant of which cannot yet be imagined in all their concreteness.

Thus, the effectiveness of strategic decisions on Russia's national development depends largely on their conception, on the correct formulation of the national mission, vision of the future, strategic priorities, and goals.

3.2 STRATEGIC GOAL-SETTING AND PLANNING TOOLS

3.2.1 THREE APPROACHES OF STRATEGIC THINKING

As V.L. Kvint writes, "There are three approaches to strategic thinking. The first approach will be referred to as the *New Horizon Strategy*. This approach requires prospective long-term thinking far beyond the current agenda of the strategic analysis object, and the ability to recognize and analyze innovative radical asymmetric and exponential pathways to success, even if they fundamentally alter the object's current activity."[13]

This is the approach the theory of noonomy suggests. From the perspective of noonomy and strategy theory, the long-term development of emerging market economies should be based on a radical departure from the observed development paradigm through profound changes in the structure and scientific and technological base of the economy, assimilation of the achievements of the new (sixth) technological mode, and a significant increase in knowledge about the directions of production intensification.

"The second approach is called an *improvement strategy*. This approach, in contrast to the former, is based primarily on a systemic analysis of the

---

[13] *Ibid.*, 32.

subsystems of the object of strategy, its elements and functions, and their interaction with each other.[14]

The choice of the first approach as a baseline does not preclude using the second approach's tools for improving the performance of public subsystems, the priority renewal of which is important as part of strategy implementation but not a strategic priority.

"The third approach can be called a *combination strategy*. This approach assumes that, in parallel to the introduction and absorption of revolutionary innovative ideas and technologies, ongoing efficiency and profitability are achieved at the expense of long-standing production and technological systems."[15] From a noonomy perspective, such an approach will not produce the desired results and will not lead to the achievement of national goals and priorities. This approach to strategy can only be applied to selected sub-systems that cannot be sufficiently resourced to bring about disruptive and innovative change.

One should note that "although quantitative analysis is fundamental, especially when assessing the resource endowment of the strategy developed through the factor of time, intuition is one of the key elements in strategic planning and strategy—called the "thinking" of the world."[16] The importance of intuition increases with a long-range strategic horizon based on the theory of noonomy. In this case, strategic calculations cannot provide sufficient and reliable empirical data and have to rely on the creativity of the human intellect.

When studying and strategizing the future, one must face the manifestations of the irrationality of a distant perspective. Irrational, sometimes subconscious and intuitive, characteristics of future periods become, in some cases, an almost unpredictable and weakly strategic unlikely the reality of the future.

3.2.2 RULES OF STRATEGIC THINKING

As V. L. Kvint writes, "Common sense, based on fleeting perceptions of reality, tends to be in direct opposition to prediction, foresight, and strategic astuteness. The strategy has to go much further and deeper than what

---

[14] *Ibid.*

[15] *Ibid.*

[16] *Ibid.*

is obvious to everyone. The strategy aims to effectively move the object of strategy toward a reality that does not exist and will only begin to take shape by the period defined by the horizon of the strategy."[17]

The theory of noonomy presents at least three successive frontiers of future reality, only the closest of which is reflected to some extent in everyday consciousness and to a large extent as something difficult to realize. It could, for example, be about using the ideas of noonomy in the processes of reindustrialization based on the latest technologies, then about building a holistic NIS.2. Finally, it could be about moving out of economic reality into the reality of noonomy. These realities do not yet exist, but the movement toward them is embedded in the contradictions of the present.

"Indeed, most have collective knowledge but lack the ability and foresight to separate and to extract "diamonds," or a truthful and visionary strategy for the future, from the tons of wastage, or primitive ideas, about the path to future success and unexpectable victories for competitors."[18] Based on the theory of noonomy, strategic thinking about social development needs to be at least decades ahead of common perceptions, going back more than a generation.

"While strategies should not rely heavily on extrapolating current or past regularities and axioms, it is even worse when they play on the lessons of history. Given the winning strategies and assumptions of the past, they should be reviewed and analyzed in the context of emerging trends, innovations and technological developments, opportunities and threats."[19]

These approaches can be summarized briefly in the seven Rules for Strategic Thinking:

Rule 1. You can't rely on common sense alone for strategy.
Rule 2. In strategy, the majority opinion is usually wrong.
Rule 3. In strategy, the present is already the past.
Rule 4. The strategist should learn and use the experience of successfully implemented winning strategies.
Rule 5. No strategy is implemented forever.
Rule 6. Cognitive inertia is the main enemy of strategic thinking.

---

[17] *Ibid.*, 34.

[18] *Ibid.*

[19] *Ibid.*

Rule 7. Strategists should not develop predictable models and scenarios for the strategy."[20]

The theory of noonomy seeks to take full account of the lessons of the past. It sees history as both an example of successful breakthroughs in national development and the root of profound crises that have afflicted different societies. These successes and failures provide material for drawing conclusions about the laws of social development and the nature of the contradictions that beset society. For example, the historical experience of both rapid industrialization and deep de-industrialization shows, on the one hand, how to achieve a scientific and technological breakthrough, and on the other, the dangers of societal degradation.

It is challenging to reach a public consensus on preparing a new strategy when the previous strategy has made the target a winner and continues to reap the benefits of success, even if strategic analysis shows a significant change in the external environment and the emergence of fundamentally new opportunities and threats. Here common sense suggests a false proposition: If the old strategy works, why improve or replace it? For example, it is invariably easier for a newly established or unsuccessful company to move to a new strategy than for a leading company to ensure that the previous strategy's harvest day is nearing its end.

Strategy making, based on the theory of noonomy, enables a change in strategy to be identified in advance if the objectives are successfully achieved. The reindustrialization of Russia based on cutting-edge technology will not allow us to be complacent because such reindustrialization creates problems as well as successes. The need to tackle them calls for a new strategy because although breakthroughs in new technologies are an absolute prerequisite for moving forward, they must be complemented by the solution of several issues relating to the social fabric, especially as these issues are caused by technological advances and the accumulation and application of new knowledge.

When an object (an organization, region, or nation) has no strategy, it is inevitably subdued by inertia. Inertia is a major obstacle to innovative strategic ideas. The larger the strategizing object, the harder it is to overcome inertia. It is therefore always easier for individual entrepreneurs, small and medium enterprises (SMEs), and small military units to shift their development vector and the overall kinematics of their activities and

---

[20] *Ibid.*, 34, 36.

to implement asymmetric export strategies in new configurations that are difficult for competitors and adversaries to predict.

The concept of noonomy implies the development of a strategy on a global or national scale; in this context, we consider the strategy of Russia's national development. The issue of overcoming the long-standing inertia of the Russian economy, which has not been able to overcome the effects of post-Soviet de-industrialization for decades, is therefore particularly acute. Perhaps for this reason, strategic decisions should initially be implemented at local points, which will become drivers of change throughout the national body.

Obviously, the implementation of any strategy will be resisted by forces opposing its goals, so it is very hazardous to propose easily predictable scenarios in a strategy. "Opponents easily strategize their consequences and implement more efficient scenarios, primarily aimed at saving time in achieving their priorities. Unconventional approaches are often the most effective ways to succeed."[21] As a rule, when an unconventional, unexpected strategy succeeds, it becomes widespread and competitors exploit it. Policymakers should be prepared for their successful strategic doctrines to be adapted by competitors and adversaries.

It is notable that the strategy of Russia's national development, based on the theoretical concept of noonomy, is to a certain extent protected from the copying of successful, nontrivial solutions by opponents or competitors. The fact is that not only do some unexpected solutions emerge from the theory of noonomy, but the very criteria for success or failure of a strategy become rather unfamiliar. The effectiveness of the solutions may therefore not be obvious to an outside observer, preventing them from assessing the potential of the strategy, and Russia from gaining a temporary competitive advantage in socioeconomic and societal development.

3.2.3 THE CONCEPT OF A HOLISTIC NATIONAL STRATEGY

"Strategy is a fundamental science, but its basic laws, principles, and categories are still evolving. Regardless of the object of strategizing, any strategy has a common nature and, therefore, should have a common theoretical basis. More detailed and specific practical recommendations require more connection with the specific characteristics of the strategic object. The practice of strategy requires methodological frameworks and

---

[21] *Ibid.*, 36.

methodological recommendations for strategies of different types and horizons.

Strategy is a systemic, multidisciplinary phenomenon. It is hierarchical in its influence, multidimensionality, and structure. Therefore, a holistic strategy system should integrate national, regional, sectoral, and corporate strategies. Moreover, it includes strategies for solving global problems, and even strategies of groups, collectives, and individuals. All these types and levels of strategies interact with one another, exerting the mutual influence of different kinematics."[22] It is easy to see that the development of the theory of noonomy alone is not enough to establish a coherent national development strategy for the foreseeable future. It can only provide a scientific platform for the development of practical solutions. A set of strategic decisions with different time horizons and varying degrees of detail is needed, based on a general theoretical understanding of the evolutionary paths of civilization and the principles of strategic planning. It should encompass not only a country's national development but also the development of various sub-systems of society, varying in role and scale, and consider the strategies and realities of external development concerning any country in the world, regardless of its territorial and economic size. Such a strategy should therefore consider the patterns of global development and the changing place of the state in a changing world and at the same time allow everyone to fit their individual strategy into the fulfilment of a national mission.

"A strategy based on the theory and methodology of strategy, and a deep and intensive analysis, can save a site from disorganization, loss of reputation, and decline, and lead it to the top of its field. However, one wrong strategic idea can ruin an entire strategic doctrine, the object of strategy as a whole, and undermine everything that has already been achieved during the development and implementation of a given strategic scenario."[23]

By proposing to look at the possibilities of a strategy based on the theory of noonomy, we are aware of the responsibility to choose the path of Russia's future development. This very responsibility compels us to set the frontiers that must be achieved to ensure that all our citizens could take their place in life with dignity. Russia needs to pull itself out of its deindustrialized state and pave the way to a possible and desirable future,

---

[22] *Ibid.*, 44.

[23] *Ibid.*

movement toward which is necessary to overcome growing problems and threats, not just for Russia but for all of humanity.

"Strategy is the product of a multiplication of time, cost, and space, where space can be understood as both developed and implemented innovative strategic ideas. In this equation, above all, two factors—time and innovation—give the strategic object a winning and hard-to-predict characteristic for competitors' acceleration and asymmetry."[24]

The societal development projected based on the theory of noonomy presupposes those necessary components of strategic decisions that derive from the theory of strategy. The whole idea of the noonomy movement is steeped in a spirit of innovation, aimed at the imperative to acquire and apply new knowledge and accelerate its transformation into state-of-the-art technology.

"Accelerating acceleration" is the leitmotif for organizing the innovation process in terms of meeting the challenge of reindustrialization and moving toward NIS.2.

The implementation of a national development strategy crucially depends on an engagement with the global development environment. This interaction takes place not only cross-nationally but also at the corporate level. The activities of large corporations have long since become transnational. "Globalization has led to corporate strategists developing strategies not only internationally but also globally. In addition, international and even medium-sized corporations are developing strategies for operating in regional economic blocs (international regional strategies) [...] The success of the corporate global strategy is linked both to reflecting the growing influence of business cultures and religious traditions of the global emerging market (GEM) in its systemic transformation and to the convergence of GEM corporate cultures with the well-established business ethos and business practices of developed-country companies."[25]

A holistic concept of national development strategy is needed to bring together all the considered levels of national development strategizing. *The mission, vision* (including the principles and priorities provided by competitive advantage) and, implicitly, *goals, placed on a timeline*, together constitute a *concept of strategy*.[26] "The three distinct categories

---

[24] *Ibid.*, 46.

[25] *Ibid.*

[26] *Ibid.*

of policy, strategy, and tactics are interrelated aspects of strategic management and governance. Their differences are as follows: when the strategy is approved and adopted for implementation, its implementation becomes a practical guide, a "guide" to the strategic object. Tactics, on the other hand, dictate daily, monthly, and annual (ongoing) plans and activities to address and resource the strategic objectives. The policy is the aggregation and integration of strategy and tactics into a single, well-functioning system. In other words: *Strategy plus Tactics equals Policy.*"[27]

> *"Strategy is a guide to set priorities and goals through the chaos of the future and the unknown. It is wisdom multiplied by a precisely chosen vector of attack with an assessment of resource constraints."*[28]

> *"Resource needs to be understood very broadly: from basic economic factors and time constraints to the impact of natural, environmental, labor, and even cultural constraints."*[29]

"When developing strategies for subordinated units, whether on the industrial plant's floor or in a country's regions, it is vital to be guided by the strategy of the integrating system within the plant or the country, respectively. Unit strategies should be more precisely adapted to their context and capacities. They may focus on specific technological, organizational, or social trends that may be too narrow for the level of corporate strategies or national strategies (when developing regional strategies)."[30]

The same applies to national strategies, where national governments issue strategic development assignments to regional and ministerial leaders without an integrated methodology. In this case, tons of wastepaper are produced with recommendations that are not only counterproductive but can actually harm the economy and, more dangerously, the national interest.

In strategy, basic economic laws, and their categories of supply and demand, and value and price, can change fundamentally as a function of time. Using time as a determining factor in strategic decision-making helps to stay ahead of competitors and adversaries, to be the first to occupy promising niches and the first to abandon unprofitable and declining ones, to be the first to exploit innovations and match their exponential nature of

---

[27] *Ibid.*, 46–47.

[28] *Ibid.*, 8.

[29] *Ibid.*, 53.

[30] *Ibid.*, 47.

emergence and development. For these reasons, the *first law of strategy* is the *Law of Time Conservation*.

The second law is the *Law of Achieving Only and Exclusively Strategic Priorities Secured by Competitive Advantages*. One of the basic principles of a strategy implemented in a competitive environment is its secrecy and invisibility to competitors and adversaries.

A strategy without tactics will not be successful enough above all on the time factor. At best, implementation will be slow, reducing the effectiveness of the strategy and allowing competitors and adversaries to implement countermeasures.

Tactics without strategy are likely to lead a strategist guided solely by tactical considerations to disaster or, at best, to a loss of competitiveness and strategic advantage.

The quest for development, long-term success, victory in competition, rivalry, and ultimately inevitable war or security requires a hierarchy of interests, a system for prioritizing competitive advantages, opportunities, goals, and an assessment of the resources required for their effective realization, considering the factor of time. This is the essence and most basic element of strategy.

The process of development, long-term implementation, monitoring, and subsequent refinements and updates of the strategy is strategy-making."[31]

3.2.4 INTERRELATION OF FORESIGHT, FORECASTING, STRATEGIZING, AND PLANNING PROCESSES

As V.L. Kvint illustrates, "A common misconception in strategy theory and practice is that the processes of forecasting, strategizing, and planning are essentially identical. Even among professional economists, forecasters and planners, these terms are often used synonymously. This "fusion" is fundamentally flawed: the three terms refer to unique professional activities that produce final products with very different intrinsic characteristics. The strategizing process ends with a newly developed and potentially implemented strategy. Forecasting employing calculations and expert assessments leads to different types of detailed forecasts. Planning, on the other hand, is an inherently different phenomenon: it focuses on management processes and results in strategic abstracts and then plans of varying levels of detail (depending on the planning horizon)."[32]

---

[31] *Ibid.*, 49–50.

[32] *Ibid.*, 52.

The process of strategy development should ensure and make use of the relationship between foresight, forecasting, strategy, and long-term planning (Fig. 3.1).

As abstract categories, visions of the future are well-established categories of scientific research. While intuition is used in strategy development and should not be neglected, it cannot be the only tool for charting a vector into the future. Crucial for strategy is the tradition of prophets and seers to associate prophecy with the time scale, as the time factor in strategy is determinative. On the other hand, philosophers do not enjoy this visionary correlation between proclaimed future processes and events and time.

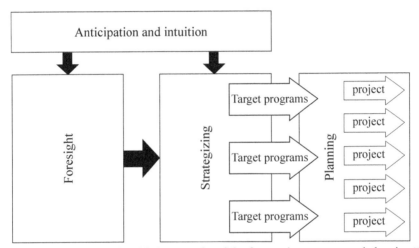

FIGURE 3.1 The relationship between foresight, forecasting, strategy and planning.[33]

3.2.5 STAGES OF STRATEGY DEVELOPMENT

Strategy development begins with an analysis of the projections, examining the external and internal environment of the subject of the strategy (Fig. 3.2). "The result of the analysis of global, regional, and sectoral trends and patterns is then used to update the global prognosis and subsequently for regional and sectoral search and target-oriented forecasts. *This creates a platform for subsequent assessments of possible competitive advantages and the selection of priorities for the strategic target based on these assessments.*"[34]

---

[33] *Ibid.*, 54.

[34] *Ibid.*, 55.

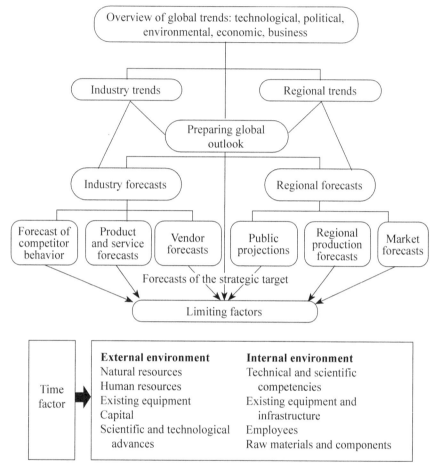

FIGURE 3.2 Stages of strategy development: study and forecasting of internal and external environment.[35]

The most valuable function of target-oriented industry forecasts is to project the activities of major competitors and adversaries and to monitor technological trends more closely.

From the perspective of the theory of noonomy, the information-technological transformation of a noonomic society should focus on the evolution of the technological structure, and the change in the dynamism of technological modes, especially in the material production sector of Industry 4.0.

---

[35] *Ibid.*, 56.

This industry is based on the robotization of technological processes using artificial intelligence and the industrial internet of things. These trends influence the informational-technological transformation of society, which in the third decade of the twenty-first century will be in the high phase of maturity and determine the main competitive advantages of countries and leaders in the global market under the long-term conditions of noonomy.

"*Regional forecasts* usually start with a search for social and political dynamics. [...] Regional dynamics must be analyzed in a cross-sectoral manner. Targeted programs should focus on trends that can lead to significant changes in the sectoral structure of the region."[36] The main focus of regional search forecasts is the monitoring of emerging market trends.

"The end result of a foresight activity is a new forecast, specific to a particular strategic target, which reveals the most important, directly, or indirectly related global, sectoral and regional trends that affect or may affect the current and future activities of the target."[37]

"A key step in the foresight phase is to scan the external and internal environment and create their objective, strategically oriented characteristics, containing assessments of new opportunities and threats."[38] The analysis of the external environment of the national socioeconomic system includes an examination of the availability of actual and potential sources of natural resources in the global economy, external labor, and capital flows. The most important external environmental factor that should be scanned and analyzed is innovative technological and scientific achievements.

Once the global, sectoral, and regional forecasts have been updated, it is necessary to "initiate the development by forecasters of an object-specific strategic forecast. This step focuses on scanning and analyzing the internal environment of the object. The competitive advantages of the strategic target in terms of exploiting new opportunities identified in the environmental scanning process and neutralizing threats and hazards are identified and highlighted."[39]

The next step in analyzing the internal environment "is to assess the technological and scientific resources of the strategic target, their relationship to emerging technological trends on the horizon and what these trends

---

[36] Kvint, V. L.; Strategy Development: Monitoring and Forecasting of Internal and External Environment; *Management Consulting* 2015, No. 7(79), 7.

[37] Kvint, V. L. *The Concept of Strategizing*, RAS-HSIU, 2019, 57.

[38] *Ibid.*

[39] *Ibid.*, 57–58.

mean for the future target in terms of, firstly, opportunities and, secondly, threats"[40] as well as the education and training of those employed in the national economy to exploit the new scientific and technological advantages (Fig. 3.3). The primary objective at this stage is to reveal the facility's unique information technology advantages, allowing it to outperform competitors, opponents, or beat them in time competition,[41] the importance of which is qualitatively increasing with the move toward noonomy. It is therefore essential to emphasize that new economic growth factors are increasing the importance of the time factor in improving the efficiency of achieving the target results.

"The next basic economic factor to be assessed is the existing production capacity and functioning infrastructure: science and technology, training, production, and social. This should also be done in the context of new opportunities in the external environment, upcoming trends, and changes."[42]

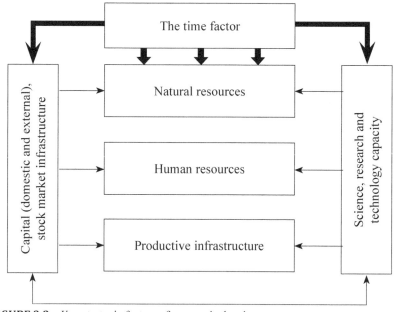

FIGURE 3.3 Key strategic factors of economic development.[43]

---

[40] *Ibid.*, 58.

[41] *Ibid.*

[42] *Ibid.*, 58.

[43] *Ibid.*

In opposition to the analysis of the internal environment of the national socioeconomic system subordinated subsystems (regions, industries, economic sectors, and companies), the scanning and characterization of the internal environment of the national system in the context of moving toward noonomy should include an assessment of natural resources and the environmental limits of their use and reproduction. This step should continue with the forecasting and characterization of the workforce and existing production capacity and capital. "Capital, employees (human capital) and, finally, raw materials (natural resources) and components used by the facility are the last basic economic factors to be evaluated for the strategic period. This scanning order shows that the *most important issue in the site-specific forecast is technology*, which can affect all other underlying economic factors of the internal environment."[44]

Defining a strategy for the national development of society under the conditions of the informational and technological transformation of its economy requires a thorough revision of all the mentioned external and internal factors, and their interaction as elements of the transition mechanism to NIS.2 and noonomy. The country's technological capabilities and, first and foremost, the ability to develop the national scientific and technological core of the economy will determine the degree of our dependence on external factors. Technological re-equipment of material industrial production will create demand for modern high-tech machinery and equipment, and the need to master their production leads to a demand for research and development. Increased productivity in high-tech manufacturing will, on the one hand, create a demand for the training of highly skilled and more culturally developed people, and on the other hand, reduce the need for low-skilled labor in the mini-grants sector. Existing development constraints on energy and transport infrastructure will generate demand for investment in these sectors.

Such an approach would require a rigorous assessment of our ability to support scientific and technological research and development, generate innovative demand, build capacity in hightech manufacturing, train human resources for high-tech manufacturing (including their cultural and creative development), expand capital expenditure resources for infrastructure development, and balance domestic production and imports of high-tech machinery and equipment.

---

[44] *Ibid.*, 58–60.

3.2.6 RELATIONSHIP BETWEEN THE MISSION AND THE FOUNDATIONAL COMPONENTS OF THE VISION

"The mission statement is the first practical document of the strategy, resulting from the scanning, analysis and forecasting of the external and internal environment, the starting point in the development of the strategy itself and the main reference point for all subsequent stages of strategic planning and the documents reflecting these stages.

In just a few sentences, the mission should be articulated in the following ways:

1. in which area the facility operates;
2. in what region;
3. and how the functioning of the socioeconomic system should benefit the population, i.e., how the given object of the strategy is unique and valuable for the consumer."[45]

The last point is key in defining the mission as the first element of the national development strategy.

The most general description of this mission in terms of the theory of noonomy was given in the first chapter of this book: to create conditions in which the growth of the human personality becomes the main result and factor of development. The mission is expressed more concretely in terms of interests, values, and development priorities.

"Since the formation of the global marketplace and the subsequent powerful wave of increased international competition and the reformatting of the world order, the importance of the mission has increased significantly: it is the first message of the facility to the outside world."[46] By declaring its mission, the Russian Federation is positioning its development strategy for all humanity in a certain way.

"The theory of strategy proves that after the mission, the vision of the strategic object (company, military unit, region, country) is always developed. The vision is one of the most misunderstood elements of strategy development. The vision is often referred to as the philosophy of the strategy in question. It would not be correct to state that it is only at

---

[45] *Ibid.*, 61.

[46] *Ibid.*

this stage that the links between the fundamental categories of strategy—*values, interests, and priorities of the* strategic object—start to form"[47] (Fig. 3.4).

From the outset of strategy development, exploring the links between the three key elements of the strategy should be the focus of attention. But the final expression of these links is in the analysis, comprehension, and formation of a strategic vision as the "Kvintessence" of strategy. The vision should not include quantitative assessments. The vision is the best element of a strategy for a concise justification of social interests, values, and development priorities.[48]

"The framing process commences with the collection of information on values, followed by the transformation of these values into interests, which are then shaped and reflected in certain priorities. Priorities are the product of the vision. They are a concentration of values and interests. In fact, priorities are the consolidating epicenter of the final version of the strategy, where all the strategy's practical implications are made explicit."[49]

The priorities in implementing the mission statement above are the strategic directions for development necessary for its achievement. Thus, from the perspective of noonomy, the strategic priorities are all shifts in technology and the social structure of production that ensure the development of a "cultural human." This means creating the conditions for acquiring a wealth of knowledge and culture, applying it in the process of creative activity and shaping a reasonable attitude of people to their needs, the interests of other people, and the state of their habitat (including the biogeosphere and technosphere).

People influenced by the strategy implementation process need to be made more or less aware of how their interests are represented and implemented through the strategic priorities. Nevertheless, these priorities can reflect only those interests that can be successfully implemented according to the adopted strategic scenario.

---

[47] *Ibid.*

[48] *Ibid.*, 62.

[49] *Ibid.*

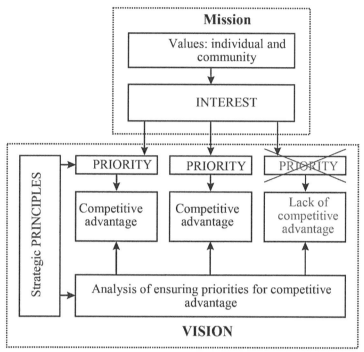

FIGURE 3.4 Relationship between the mission and the core components of the vision.[50]

"A strategy almost always implies the implementation of more than one priority, especially when it concerns global or large-scale interests. The first strategy implementation period narrative includes—only and exclusively—priorities secured by effectively implemented and innovative competitive advantages. The first period also finances the creation of new—or the regeneration of previously lost—competitive scientific, technological, and other innovative advantages and the training of highly qualified, highly specialized personnel to implement the priorities of subsequent strategic periods."[51]

"It is vital to underscore that strategies are developed to achieve priorities, not to solve any problems that the subject of the strategy may be facing. Only this strategic approach can ensure long-term strategic success. Problems are often fleeting: they come and go like the morning fog; priorities

---

[50] *Ibid.*, 63.

[51] *Ibid.*, 64.

reflect the long-term vision of the strategy and the underlying interests and values of the people concerned."[52] The surest way to achieve even the most complex strategic directions is to set strategic priorities.

"The exact definition of priorities is fundamental because all the resources of the object of strategy are concentrated around them. Equally important, priorities are the guiding pathways for selecting goals and then objectives for the object of strategic planning."[53]

3.2.7 GOAL-SETTING AND TARGETED PROGRAMS

"Goal-setting is the beginning of turning strategy into a practical reality. Based on the mission statement and the philosophical foundation of the vision, goal-setting is the qualitative orientation, specification of priorities, subordination, and interrelationship of the individual goals of the object strategy. The length of the goal-setting section and the formulation of strategy objectives can vary from a simple listing of them to several dozen pages with a detailed justification and description of all their detailed qualitative characteristics."[54]

While the number of priorities is usually limited to a few items (as their excessive number leads not just to the scattering of resources but the vagueness of the very vision and mission of the strategy), the goals of strategic projects are more diverse and may involve intermediate goals (subgoals) and stages (sub-stages) of their implementation, forming a rather complex system.

"The objectives must have clear qualitative criteria for their achievement since, in general, the objectives are transitions of certain elements of the socioeconomic system of society and its material basis into a qualitatively new state.

To implement each strategy goal, a target program is developed that concentrates time-bound and fully resourced objectives. If the goal is a qualitative orientation of strategic priority implementation, then the objectives are the first element of the strategy, having reasonable quantitative characteristics.

---

[52] *Ibid.*

[53] *Ibid.*

[54] *Ibid.*, 65.

The mission, vision, goals, and objectives of the strategy are not only the stages of its development but also the formalized, independent, and at the same time interrelated and mutually complementary, main elements (documents) of the strategy.

Defining objectives is the first stage of strategic planning, where quantitative characteristics and evaluation indicators are established.

Given the content of all the previous stages of development and strategy formation, the implementation of the tasks is closely linked to the time scale. This is, in fact, the main timetable for the further development and implementation of the strategy.

All strategy objectives are set in the context of the resource limits of the object under the determining influence of the time factor.

When the most decisive organizational forms and elements of the strategy (reflected in documents)—mission, vision, goals, target programs, strategic projects, and technology platforms integrating the resource objectives—have already been developed and pre-approved, the next step in the strategic planning process is to select at least three possible and necessarily alternative strategic scenarios through which the approved priorities, goals, objectives are achieved. The strategic plan for implementation, projects, and platforms are developed once a preferred scenario has been selected."[55]

"The most important constraints, apart from time, are those of the external environment, as they are independent of the object of the strategy or very limited in their influence in contrast to internal resources, which can be changed if this becomes necessary.

The external environment concerning a facility represents a systemic risk to its strategy. It is possible to predict and forecast systemic risk, although it is practically impossible to manage it (or possible, but within minimal limits).

The strategic plan differs significantly from the current (annual) or operational plan. The strategic plan needs to be blockbased, more aggregated, and more flexible, allowing for adaptation to future difficult-to-predict conditions. The strategic plan includes the elaboration of specific constraints on the future functioning of the site, analyzing the five basic economic factors. This analyzes the scale needed to achieve the goals and objectives of the strategy, its projects, raw materials and components, human resources, investment and operating capital (in the form of a large-scale budget), production capacity and other infrastructure."[56]

---

[55] *Ibid.*, 65–66.

[56] *Ibid.*, 67.

The Basis for Strategizing National Development

Figure 3.5 shows the "interconnection of all the main strategic documents, from the mission formulation stage to the development of the strategic scenario and then to the strategic plan. When strategists present the strategy, all these documents should be made available to all authorized representatives of clients with access to confidential information, including executives, commanders, and members of key collaborative bodies, governments, and so on.

Realistic time constraints should be set for all phases of the strategy, from entry strategy to exit strategy, including all underlying economic factors, their allocation and placement, and their combination for use in strategy implementation processes."[57]

FIGURE 3.5 Shaping strategy.[58,71]

---

[57] *Ibid.*, 67–69.

[58] *Ibid.*, 68.

"The professional strategist seeks and justifies new strategic avenues, sections priorities, and develops scenarios in an environment where the past is only partially extrapolated into the future, the 'present' does not exist, and future social processes and economic agents remain largely unknown, even to strategists with a long-term vision."[59]

---

[59] *Ibid.*, 20

CHAPTER 4

Strategic Goals of Socioeconomic Development

4.1 IDENTIFYING STRATEGIC TARGETS

A coherent strategy system should integrate the global, national, regional, sectoral, and corporate levels. For a national strategy, the national and global levels are determinative of all others. The task of regional, sectoral, and corporate strategies is to refine strategies at the national and even global level.

The strategy concept includes a mission statement, a vision of the future, and development goals placed on a timeline. There is no strategy without an understanding of the chronological framework for its implementation. The specific chronological framework of the strategy cannot be defined a priori. It is only by setting specific objectives and plans that it is possible to quantify the chronological scale of the strategy. Yet, a qualitative assessment, i.e., the sequencing of goals and objectives, is already possible in the first phase of strategy development.

For this matter, a development forecast is drawn up to prioritize the implementation of the strategy. Of course, in prioritizing a national strategy, it is essential not only to have a forecast of the development trajectory, but also to build on the national development mission and goals that have already been formulated.

The mission positions the object of the strategy, e.g., the national socioeconomic system, to the outside world. But the mission is not only outward-oriented. It reflects national values, interests, and priorities, and serves as a platform to consolidate society to achieve strategic goals. As part of the development strategy concept, the national development mission derives not only from the aspirations of the people living in the country, but also from objectively determined trends that put on the agenda the improvement of the quality of life and the conditions for the development of human potential.

The very movement toward ensuring human dignity can be defined as the mission underpinning the national development strategy. Living well in terms of the theory of noonomy includes not only a rational level of consumption but also the involvement of people in activities that contribute to the development of the human personality and the elimination of risks associated with the imbalance of the natural environment and human intervention in nature. Pursuing this mission is a major development strategy for our society and humanity as a whole.

The goals of the strategy determining which milestones the country should aim for are directly linked to deciding on the mission. Having defined the mission, the priorities and objectives of the strategic development need to be justified and formulated.

A mission not oriented toward achieving new horizons but only toward maintaining the status quo also requires considerable effort to ensure progressive socioeconomic development. For example, the concept of the transition to the new industrial society of the next generation and to noonomy, which is based on understanding the development of the world economy and its underlying intentions, leads to a rather strong conclusion about the gradual decline of the commodity component in the system of social production.

Russian Academy of Sciences (RAS) Academician S. Yu. Glazyev's position on this issue is unambiguous: "The demand for hydrocarbons and other modern energy sources [...] is about to decline sharply. Therefore, today the construction of gas pipelines on the seabed is madness against the background of quite obvious objective shifts in the coming structural dynamics of global energy consumption."[1]

Raw materials producers will lose their position in the global economy, and the leadership will finally be reserved for those who develop and apply advanced technologies. The economic leaders of the coming decades are technological leaders.

Moreover, it is apparent that more and more in today's world we can "speak of the potential long-term enslavement of those countries that will not possess advanced technologies in some 30 to 40 to 50 years unless they create institutions, tools that enable them to establish technological parity or achieve at least in some areas technological leadership."[2]

---

[1] Glazyev, S.Yu.; Prospects of Formation in the World of a New VI Technological Way; *MIR (Modernization. Innovation. Development)* 2010, *V. 1, No. 2(2),* 7.

[2] Bodrunov, S. D.; Coming and Thinking; *Economic Revival of Russia* 2016, *No.4 (50),* 17.

For example, this approach is relevant for Russia due to the technological gap with the most developed countries in several industries. The Russian academic community has adopted this approach. Thus, RAS Academician A.D. Nekipelov emphasizes: "The economy is bleeding, much [...] has been irrevocably lost, much is on the verge of extinction. The prospect of being on the sidelines of the global economic community for decades to come is closer than ever. That's why it's necessary to create conditions for focusing minimal resources on carefully selected areas that offer a chance to return [...] to the ranks of advanced economies for the foreseeable future."[33]

The implementation of a strategic mission requires a long-term orientation toward the most advanced milestones. Therefore, a new qualitative state of society, based on the theory of noonomy, i.e., the transition to NIS.2, should be taken as a benchmark. In such a case, the intermediate goal, without which it is impracticable to reach the NIS.2 milestones, should be the reindustrialization of Russia on a state-of-the-art technological basis. There is no alternative to reindustrialization for Russia. It is imperative to figure out what specific tasks need to be undertaken to achieve this goal.

It is worth noting that many countries face similar challenges related to building industrial and technological capabilities.

It is strategically necessary to orient oneself toward "technological development, accelerating progress, working for human development, not against it. And re-industrialization should be high-tech, on an advanced, knowledge-intensive technological basis."[4]

It is important to "unfold" financialization, to give it the function of productive capital. Thus, in terms of general trends, Russia has opportunities and chances for reindustrialization, as well as technological reserves in all the practically promising areas and, what is important, in material opportunities. Russia possesses sufficient scientific and industrial-technological potential and can position itself among the leaders since the entire transition phase to NIS.2 creates the preconditions for Russia's dynamic development if resources are allocated to strategically justified priorities. It follows that "we must work toward a maximum convergence of the components of the industrial noo-industrial process, shortening the path

---

[3] Nekipelov, A. D.; Globalization and Strategy of Russia's Economic Development; *Problems of Forecasting* 2001, *No. 4*, 12.

[4] Bodrunov, S. D.; On the Issue of Noonomy; *Economic Revival of Russia* 2019, *No. 1(59)*, 8.

from knowledge to product, incorporating knowledge into both product and skills/competences, creating, through the integration of production, science, and education, industrial complexes, new types of industrial sector entities, which will replace in the future the current traditional type of production."[5]

It is not only with digitalization that we can achieve the crest of the science and technology wave. What is needed is not the "digitalization of backwardness" but a qualitative leap in productivity, which requires both a technological overhaul of the entire complex of industries and a concomitant change in socioeconomic institutions.

It is critical to appreciate that the strategy should focus not on arithmetical growth in GDP, profits, consumption, etc., but on qualitatively new development. It is strategically necessary to improve the quality of the technology base. Therefore, it is important to raise the level of satisfaction of real needs, not to operate in a paradigm of "more and more," and not to flood the market with simulacrums, surrogates, and obsolete signs of needs satisfaction. This approach is explicitly dictated by the increasing share of knowledge in all components of production.

It goes without saying that to reequip fixed assets with qualitatively new technology, the production of new, modern equipment must be increased. But what matters is not the volume but the ability to produce what is at the cutting edge of the world's modern knowledge.

The same goes for consumption. The focus on quantitative growth should be shifted to long-term goals that determine the level of rational needs to be met. It is therefore vital, for example, to ensure the quality of food, its storage and processing, and the quality of transport links. In the years ahead, the main parameter will be a qualitative improvement in the level of comfort of the living environment.

Thus, from the height of the world's knowledge and the ability to implement this knowledge into new technologies, the use of knowledge must now be oriented toward creating a means of production that meets needs on a qualitative rather than a quantitative basis, especially in terms of GDP, which does not even do a good job of capturing growth in volume. It is important to formulate a strategy that continuously generates new technologies in the process of transformation to NIS.2. This approach will

---

[5] *Ibid.*, 16.

ensure not inertial self-development but the creation of mechanisms of self-support and qualitative change.

To this end, it is necessary to improve the mechanism of public administration. RAS Academician V. Kuleshov comes to the following conclusion: "Prolonged stagnation, degradation of entire economic sectors is gradually transforming from an economic problem into a social and political one. It is clear that the current model must be changed. However, the predominantly market-oriented methods that have worked well in other socioeconomic realities, which have been tried and tested, do not meet this challenge. The creative component needs to be strengthened, and the role of the state in the transition to a new (innovative) form of development needs to be increased."[6]

Thus, the economic community faces the challenge of finding *a new model of economic development and growth*, and more broadly, a *new economic doctrine*.

Despite their huge scientific potential and excellent achievements in several knowledge and technology fields, countries that rely on commodity exports generally lag significantly behind the more advanced economies in terms of technology. In some sectors, however, these countries have been able to maintain or renew advanced levels of technology. Such countries must create and harness economic forms that enable a transition to a new level of material production: NIS.2. Otherwise, there is a growing risk of the objective needs development de-synchronization for advancing cutting-edge technology on the one hand, and the formation of the social relations and institutions that must meet those needs on the other.

The new economy must be built on a narrow segment, practically working out what will inevitably become the future of economic activity and societal development.

Consequently, resources need to be identified, as well as ways to achieve these goals.

Re-industrialization requires a qualitatively new set of targets, without which re-industrialization will remain a pious wish. Changes will be required in all the components that determine the technological level and the economic and institutional set-up of social production.

---

[6] Kuleshov, V. V.; Alexeev, A. V.; Yagolnitser, M. A.; InterExpo GEO-Siberia, XIV International Scientific Congress, April 23–27, 2018; Proceedings in 2 vol.; SGUGiT, 2018, V. 1, 298.

4.2 NATIONAL PROJECTS

In Russia, the most important, encompassing areas of development are targeted programs that often take the form of national projects. From the strategic vision perspective, national projects in this form are a set of insufficiently linked tasks that are not subordinated to the mission and objectives of the national strategy. The problem, however, is that in Russia as of 2020, these core elements of the strategy are not explicitly defined and cannot serve as the basis for the development of targeted programs. However, the programs themselves are often not adequately quantified and are not always adequately resourced. Such national projects cannot be a decisive element of a coherent strategy.

The weaknesses of the national projects are also noted by RAS Corresponding Member A. A. Shirov: "Unfortunately, the issues of modernization of the basic sectors of the economy are not directly addressed by the national projects. They talk about building infrastructure, investing in human capital, but the mechanisms of how the industrial core, which generates the primary income without which it is difficult to ensure the mechanisms of expanded reproduction, will be modernized, have so far been overshadowed."[7]

However, national projects can be an effective strategic element of a coherent national strategy. The solution lies in making national projects subordinate elements of the national strategy, stemming from the definition of the national mission, strategy objectives, national priorities, and the range of tasks needed to achieve these priorities.

For example, if the strategy identifies digitalization and the development of the digital economy as an important priority, then the implementation mechanism of the strategy should develop a resource-intensive target program. The program integrates all subprograms, combining objective-oriented tasks. This is a big, serious, and complex job, which can only be implemented based on a strategy and subsequent strategic plans. It should include strategic planning, selective planning in terms of choice of directions, indicative planning in terms of measuring results, etc. This type of planning should be mandatory, as it establishes both "pathfinders" for business and "benchmarks" for the public administration system.

---

[7] Shirov, A. A.; Problems Of Reproduction in The Modern Russian Economy; *Voprosy Politicheskoy Ekonomiki* 2019, *No. 2*, 45–46.

For this to happen, *the Russian economy needs a major systemic change:* a transition to economic management based on scientific foresight, long-term strategy, targeted programs and medium-term indicative plans implemented through a proactive industrial policy. The state should guarantee business paternalism concerning long-term investment in R&D and technological reequipment. It should ensure stable, supportive taxation and comfortable credit conditions for the real sector, especially the high-tech sector. At the same time, such a system should (and can) ensure a moderate level of social differentiation: citizens' incomes (after the deduction of minimum income provided by the state according to certain social criteria) should depend mainly on their real contribution to the economy.

High standards can only be achieved through strategy and effective planning, and combining the market and plan, building on the successful experiences of China and the Scandinavian countries.

This can be addressed technologically by applying modern information technologies, utilizing the idea of distributed databases (in its modern form as embodied in blockchain technology, for example), greatly increased computing power to optimize solutions based on very large data sets, and so on. The combination of modern information and communication systems with the capabilities of cognitive technology, artificial intelligence, self-learning systems, human-machine systems, etc. makes it possible to digitalize both planned and market-based approaches to optimizing economic decisions, and to integrate these two approaches.

Of course, the new technological base will gradually change the institutional structure of the national economy, allowing it to effectively reorient it toward reindustrialization, creating the material basis for a technological breakthrough into the future.

CHAPTER 5

Priorities for the Modernization of National Economy

5.1 THE ACCELERATION OF SCIENTIFIC AND TECHNOLOGICAL GROWTH FOR THE PURPOSE OF ACHIEVING A NEW QUALITY OF LIFE

5.1.1 THE "NEW NORMAL" AND THE NEED FOR A TECHNOLOGICAL SHIFT

The state of the global economy at the beginning of the third decade of the twenty-first century is its new "normality," shaped by the global pandemic and other processes of global significance, including the investment downturn in the global marketplace.[1]

Research shows that the global trends outlined above are the result of qualitative disparities between outdated forms and methods of economic organization and new technological possibilities. Disproportionality causes persistent market fluctuations and growing tensions in the global market space.

This is the stage of transition to a new development paradigm.[2] A crucial feature of the "new normal" is recognizing the need to ensure

---

[1] The lack of investment, which threatens the development of recession, is complained about by World Bank experts (*see*: The World Bank Pointed Out the Risks for the Global Economy; *Vesti. Economy*, January 10, 2018. http://www.vestifinance.ru/articles/96065), and OECD experts (Bazanova, E.; The Global Economy Is Trapped in Low Growth Rates OECD; *Vedomosti* 2017. https://www.vedomosti.ru/economics/articles/2017/03/09/680409-mirovaya-ekonomika-popala).

[2] *See,* for example: Bodrunov, S. D.; On Some Issues of Evolution of Economic and Social Structure of the Industrial Society of a New Generation; *Economic Revival of Russia* 2016, *No.3(49)*, 5–18.

further technological development as the basis for the entire civilization's movement.[3]

For the most advanced countries, the severity of the problem of accelerating industrial and technological development is not so apparent; it is hidden behind a higher (than in the mainstream) level of technology, greater research and development capacity, and the appearance of a continuous flow of innovation.

During the third decade of the twenty-first century, the global economy will evolve into a new technological mode, in which technological change will become an integral part of the production process. This will generate new demands for the integration of production, science, and continuing education.

In such circumstances, *Russia has a "window of opportunity"* associated with the fact that today's global capitalist economic model is stalling qualitative, revolutionary shifts of high technological significance.

5.1.2 *CONTINUITY OF THE INNOVATION PROCESS*

The "Strategy for Innovative Development" seeks to respond to the challenges and threats faced by Russia in the field of innovative development and to define the goals, priorities, and instruments of state innovation policy.[4]

To facilitate the generation, selection, development, and translation of new ideas into innovative technologies, special institutions are required, which in world practice are called *national innovation systems (NIS)*.

An innovation system is usually defined as *"an organizational and economic mechanism* with an appropriate *infrastructure*, orienting scientific organizations toward achieving commercial and social benefits from

---

[3] *See*, for example: Bodrunov, S. D.; Modernization of the Defense-industrial Complex and Ensuring State Security; *Year of the Planet* 2005, *No. 14*, 107–112; Bodrunov, S. D.; *Analysis of the State of Domestic Machine-building and the Imperatives of New Industrial Development*; Institute for New Industrial Development (INID), 2012; Bodrunov, S. D.; *New Industrial Development of Russia in the WTO Environment: Examination of the Adopted Concepts of Innovative Development of Russia*; Institute for New Industrial Development (INID), 2012; Bodrunov, S. D.; On the Issue of Reindustrialisation of the Russian Economy; *Economic Revival of Russia* 2013, 4 (38); Bodrunov, S. D.; Russian Economic System: The Future of High-tech Material Production; *Economic Revival of Russia* 2014, No. 2.

[4] Bodrunov, S. D.; Lopatin, V. N.; Strategy and Policy of Reindustrialisation for Innovative Development of Russia; Institute for New Industrial Development (INID) 2014, 43.

developments, production organizations toward the constant renewal of products and technologies, the organization of production, work, and management based on the use of these developments, and authorities and civil society toward the development of mass innovation activity.

The national innovation system can also be defined as a *set of individual institutions* that, collectively and individually, contribute to the development and diffusion of new technologies, which form the framework from which government formulates and implements policies to influence innovation processes."[5]

Essentially, NIS is *"a system of interconnected institutions for creating, storing and transferring knowledge, skills and abilities that define new technologies.* From this perspective, neither re-industrialization in the aforementioned interpretation, nor the further successful development of modern production (much less future production) is possible without a *deep integration* of production with education and science both as ideology and as a result of it. *The integration of science, production and education into a single system* is *a necessary organizational condition* and *a prerequisite for the practical realization* of reindustrialization in the Russian economy."[6]

The Achilles' heel of the Russian innovation system is its *low performance*, which is caused by "blurred interests of NIS participants, inconsistency, non-interdependence of these interests, lack of proper economic motivation, a non-harmonized system of indicators of innovation performance for different NIS entities, an underdeveloped intellectual property market, and so on."[7]

At this stage, despite gradual progress in solving the problems of import substitution, given the level of scientific and technological results in different countries, a full-scale technological modernization of the Russian economy is impossible without a significant *transfer* of a wide range of foreign technologies (for example, until 2014 inclusive, the country purchased foreign technologies annually for 140-165 billion dollars). However, due to financial and political restrictions, Russia did not receive a complete and necessary set of imported technologies. (It should

---

[5] Bodrunov, S. D.; Innovative Development of Industry as a Basis for Technological Leadership and National Security of Russia; *Proceedings of the Free Economic Society of Russia* 2015, No. 3, V. *192*, 36.

[6] *Ibid.*, 38.

[7] *Ibid.*, 40.

be noted that these restrictions have a positive aspect, forcing the country to speed up and clarify the objectives of technological modernization.)

The most important difference between today's industrial production and the stage where a set of active industrial policy measures first became widespread and successful is its *innovative* nature, based on a *knowledge-based economy.*

This is why it is not just a question of creating a list of new technologies but of transforming the process of creating these technologies into a continuous flow.

Undoubtedly, the flow of new technologies has always existed in industrial production, regardless of the social system. However, since the late twentieth century, "the flow of innovation has become continuous, and *continuous updating* of *product* lines and the development of *new technologies have become an imperative for the efficient functioning of production.*"[8] Prospective production "acquires the character of "continuous innovation"; research, search, transfer, and implementation of technologies become integral elements of a modern production system, part of the production process."[9] And such elements of intersubjective relations between scientific and production structures in the framework of industrial activity as *technology* transfer is a mandatory element of the production process. "At the national level, the need to facilitate this flow of innovation leads to the transformation of R&D into a special (potentially vital and large) sector of the national economy and to the formation of *national innovation systems* that serve all stages of *the innovation process* in national economies."[10]

5.1.3 THE ROLE OF IMPORT SUBSTITUTION IN TECHNOLOGICAL DEVELOPMENT STRATEGY

Consistent *implementation of the import substitution strategy* helps to minimize the negative effects of economic sanctions and should be *a central element and focus of all government economic (above all industrial) policy in Russia.*[11]

---

[8] Bodrunov, S. D.; What Kind of Industrialisation Does Russia Need? *Economic Revival of Russia* 2015, *No. 2(44),* 13–14.

[9] *Ibid.,* 14.

[10] *Ibid.*

[11] Bodrunov, S. D.; *Theory and Practice of Import Substitution: Lessons and Problems,* S.Yu. Witte, 2015.

Equally essential is the development of new areas of exportable/ exportable products: machinery and equipment, technology, know-how, and educational services. An even more promising direction could be developing and implementing, together with foreign countries (including Asian and Latin American countries), long-term programs for "growing" integration structures combining production, science, and education in the so-called clusters.

It is worth noting that "import-substituting growth strategies have been used by various countries,[12] most notably those in Latin America (Brazil, Argentina, and Mexico) and Asia (South Korea and Taiwan). The instruments to stimulate import-substituting growth were as follows:

- protectionist measures and, in particular, state-subsidized price reductions for domestic products;
- restrictions on imports of industrial products from other countries;
- investment of funds *withheld* in the state from the sale of import-substituting products in the modernization of industrial enterprises.

[...] In Brazil, the import substitution policy ("Plano Brasil Maior") was originally aimed not so much at limiting imports as at stimulating exports. The program guaranteed national producers-exporters a partial tax refund and the possibility to benefit from a specially created state fund to finance export operations. The country has established internationally competitive manufacturing industries, particularly in the aircraft industry (Embraer), as well as mechanical engineering and shipbuilding. Oil producers and metallurgical companies increased their exports."[13] This has allowed Brazil to increase its economic growth rate and become one of the fastest-growing economies globally.

The experience of South Korea has also been positive, "using import substitution not as an independent growth mechanism but as a transitional policy to strengthen the national economy and build strong export capacity. Such a strategy has been called "export-oriented import substitution."[14]

---

[12] *See*, in particular: *The Third World: Half a Century Later*, Khoros, D.B.; Malysheva. M.; eds.: IMEMO RAN, 2013.

[13] Bodrunov, S. D.; Rogova, E. M.; On the Basic Principles of Import Substituting Industrial Policy Formation in Russia; *Actual Problems of Economy and Management* 2014, *Issue. 4(4)*, 8.

[14] *Ibid.*

An effective import-substitution strategy and increase in high-tech exports necessitate restoring the structure of the domestic industry, recreating basic production niches that were replaced by foreign manufacturers during deindustrialization and have led to the current setbacks.[15] Do not underestimate the complexity of this task. After all, it is not just a question of maintaining or increasing the rate of economic growth. As RAS academician A.A. Aganbegyan notes, we need not only "to ensure economic growth, but to ensure it based on modernization, i.e., technological renewal of the economy, major structural reorganization of the national economy, to get off the oil and gas needle and radically change the export structure in favor of high value-added, high-tech products in particular."[16]

Re-industrialization, implemented based on the experience described above, should lead to progressive changes in the structure of the economy, to its diversification, and to the correction of its lopsided fuel-raw orientation. These structural shifts should increase the scientific and technological independence of the Russian economy.

The latter requires a significant change in the weighing of the two groups of sectors:

- sectors of high-tech material production, providing the rest of the economy with modern machines, equipment and instruments;
- research, development, education, and health sectors.

These sectors ensure the flow and technological application of new knowledge, as well as the creation of the human capital needed to drive the economy forward.

5.1.4 TECHNOLOGICAL PROGRESS AS A BASIS FOR INCREASED NEEDS SATISFACTION

The possibilities of modern technologies allow them to be brought together to meet increasingly complex challenges and needs. The global economy has been experiencing a constant slowdown over the past 20 years. One exception is China, which has been developing less intensively for more than two decades but by extensifying large-scale industrial production.

---

[15] Bodrunov, S. D.; *Formation of Russia's Reindustrialisation Strategy, Edition 2*; Part One, INID, 2015; Part Two, INID, 2015.

[16] Aganbegyan, A. G. Human Capital and Its Main Component the Sphere of "Knowledge Economy" as The Main Source of Socioeconomic Growth; *Economic Strategies* 2017, *No. 4*, 12.

However, in terms of meeting human needs, the situation seems to be quite the opposite (contrary to traditional statistics, which do not meet real research requirements). It could be claimed that humanity is now entering a "golden age" in terms of meeting its needs.[17] A careful analysis allows us to speak about this quite definitely.

However, the growth rate of demand satisfaction is increasing faster than the growth of economic indicators that characterizee the growth of social wealth, such as GDP, and sometimes "despite" the changes in GDP.

"Consider a use-value designed to satisfy a particular need of people. Take the watch, for example. They satisfy the need to know the time. For example, some watchs cost $100 20 years ago. Now, we have mobile phones. The first phones cost about $1000. The person who bought the phone satisfied their need to communicate with the caller in mobile mode. Thus, a person who simultaneously satisfied two such needs created a demand for $1100 (for a watch and a mobile phone). However, the development of technology has led to a technological synergy. New gadgets, after a while, already contained two functions—time and mobile communication—and advances in technology made it possible to reduce the cost of producing a "single" product that already satisfied two needs/functions. Suppose such a gadget became worth $300. Thus, a person who wanted to fulfil these two needs now creates a demand for $300. That is, in terms of the statistics used by the global economy, we are seeing demand fall as it has fallen from $1100 to $300.

This will lead to a decrease in GDP in terms of standard statistical methods (Fig. 5.1). It is vital to note that the number of people who would want to satisfy the two needs for $300 is significantly higher than the number of people who would be able to satisfy them for $1100. The number of people who can afford two needs for $300 is indeed much higher than the number of people who can afford the same for $1100. However, the number of people willing to satisfy these two needs at all is limited, and the total demand for these two needs created by people in the new situation will, if this trend continues, sooner or later be less than the total demand created by the number of people who could satisfy these two needs for $1100. So, since the number of consumers is physically limited, sooner or later this trend will lead to a decrease in the statistical volume."[18]

---

[17] Bodrunov, S. D.; *Noonomy*, Cultural Revolution, 2018, 95.

[18] *Ibid.*, 95–96.

Such trends have already been noted by economists: "New technologies can be very useful for consumers but are not adequately reflected in GDP growth,"[19] corresponding member of the Russian Academy of Sciences E. Dementiev has stated.

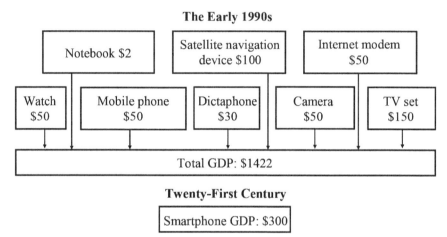

FIGURE 5.1 Synergy of meeting needs in one gadget reduces GDP (prices are notional).

Thus, we see a fundamental divergence between the "accounting" picture and the reality reflecting the actual satisfaction of needs. When you consider the myriad of combined functions to meet people's ever-increasing needs combined in new knowledge-intensive products, you get not a slowdown in economic growth but a dramatic increase in the ability to meet needs. We might claim that we are now quietly entering the age of NIS.2, which will be characterized by an increasing willingness to meet people's growing needs through advances in technology.

So, a knowledge-intensive product is evolving in its growing ability to meet an ever-wider range of human needs (the above evolution from watches and phones to smartphones with a huge increase in the range of functions). Advances in technology are making it possible to meet multiple human needs, previously met by different/few industrial products, with a single knowledge-intensive industrial product in the era of NIS.2. This is

---

[19] Dementiev, V. E.; Productivity Paradox in The Regional Dimension; *Regional Economy* 2019, *V. 15, Issue. 1*, 46.

a good illustration of the philosophical principle of reciprocal reflection in economics: all actors influence each other so that new needs develop at the expense of old ones. Technologies that are created to meet some needs simultaneously open up the possibility of meeting new ones.

The potential for far greater saturation of people's needs outstrips the growth of those needs. The interesting question is (incidentally) about the current unused potential of new products: How to utilize these potentials so that they do not go to waste?

It is evident that in knowledge-intensive production, the use of material resources per unit of "old" demand is significantly reduced while maintaining/increasing the share of knowledge in the knowledge-intensive product. This mainly reduces the cost of producing a knowledge-intensive product per unit of "old" demand. This leads to a synergistic drop in demand for traditional materials, resources etc., and a corresponding drop in the importance of raw materials for the new global industry. Russia's continued de facto apologetics for a resource-based economy (while condemning it verbally) leads to a deadlocked development strategy.

This is confirmed by the conclusions of RAS Corresponding Member D.E. Sorokin: "Forecasts of Russia's economic growth based on projected estimates of the country's reserves of mineral resources and the foreign economic conjuncture for those resources can only point to the potential instability and historical futility of the energy resources scenario."[20] Academician V.V. Kuleshov comes to the same conclusions: "Simulation modeling carried out using a cognitive model shows that the outpacing development of the mineral resource complex (extractive industries) even in a favorable external environment cannot ensure sufficient economic growth and sustainable socioeconomic development."[21]

On the other hand, "the relative reduction in the resource intensity of production that is occurring along with increased knowledge creates a platform for regulating an appropriate level of equilibrium with the natural environment and overcoming ecological problems."[22]

---

[20] Sorokin, D. E.; Conditions for Transition to Innovative Type of Economic Growth; *MIR (Modernization. Innovation. Development)* 2010, *T. 1, No. 2(2)*, 28.

[21] Kuleshov, V. V.; Alexeev, A. V.; Yagolnitser, M. A.; *Assessing the Role of Natural Resources in The Country's Economic Growth: Cognitive Analysis and Decision-Making*, Interexpo GEO-Siberia: XIV International Scientific Congress, April 23–27, 2018: Proceedings in 2 vol., SGUGiT, 2018, V. 1., 305.

[22] Bodrunov, S. D. *Noonomy*, Cultural Revolution, 2018, 98.

Nevertheless, an increase in the level of satisfaction of material needs does not solve all problems. The fundamental contradiction that has marked the entire course of human development (between the sphere of services, the sphere of the production of living conditions and human life itself, self-reproduction as an individual) remains.

The resolution of this contradiction occurs objectively, with the growth of human knowledge and the development on this basis of knowledge-intensive technologies capable of satisfying vital needs at lower and lower costs. At the same time, the share of spiritual needs increases. It is not the material conditions of existence (insofar as they are already provided for) that become the main focus of needs satisfaction—spiritual values come to the fore.

The history of humanity has seen a ripple effect on the role of spiritual values. Such a thesis echoes Lev Gumilev's theory of passionarity.

From time to time, there has been and continues to be a shift in the balance between the need to maintain the body and the need to maintain one's inner self. There have been periods when the spiritual dimension has come to the forefront for some (not all) people, such as during the times of early Christianity and the Renaissance. During the Renaissance, the development of new technological and economic systems (handicrafts, the market, etc.) stimulated a shift in the direction of spiritual needs, although in a small part of the population at first. This was reflected in new artistic techniques, new genres, the appearance of new musical instruments, and the emergence of universities.

The influence of changes in material production on the evolution of the spiritual component of needs is not of direct rigid dependence. Rather, the connection is mediated by the totality of societal conditions. Shifts in technology, resulting in changes in technological modes, are superimposed on changes in a nation's social order and structure. In fact, this is what John Kenneth Galbraith observed in his book *The New Industrial State*.

Shifts in the structure of needs and the gradual shift in the importance of spiritual needs are often reflected in a crisis of the educational system. For example, medieval universities emerged as a response to society's new spiritual demands.

The early twenty-first century has also witnessed the development of a crisis in a broken educational system. The role of self-learning is increasing, and there is an emphasis on self-education. The reason is the increasing importance of knowledge and its volume, while the capacity to acquire this knowledge is limited. It is apparent that all knowledge cannot be embraced and that individualization of education makes it possible to tailor learning and knowledge acquisition to the individual's unique characteristics and to make the learning process more effective.

This is facilitated by new means of communication, transmission, and assimilation. Individual gadgets and technical means allowing access to the virtual information space at any time are becoming widespread. New forms of communication are also emerging through the virtual space: by interacting with anonymous phenomena on the web, one essentially communicates with oneself. But even such communication can be productive if the result is a rethinking and reassessment of oneself and one's relationship to the world around us.

In *the new society, not only a new hierarchy of needs is being built, but also a new hierarchy of values*. The need for self-worth, *the need for personal development,* communication, social acceptance, and self-esteem comes to the fore.

5.1.5 INEQUALITY: NECESSARY AND UNNECESSARY. NEW INEQUALITY

The inequalities based on the tensions associated with the shift in contemporary geo-economic relations[23] in the world-system and the conflict between "catching up" and "under-achieving" economies in the global economy have been described above.

But as we move toward NIS.2, a new and different inequality will develop in the economy not in the capacity to devour natural resources in pursuit of unbridled consumption of material goods and services, but in the satisfaction of cultural/spiritual needs and opportunities for personal development, creativity, and cultural aspirations.

The same factors will determine the differences in the level of development of national economic complexes. Therefore, the future of our economy lies in the unconditional rejection of reliance on oil revenues and other natural resources as urgent "social patches." There is a need for a strategic shift in investment flows toward sectors that focus on developing human qualities.

Flat egalitarianism has never been a sustained phenomenon in history. Without a certain level of property and income inequality, there can be no incentive for economic development. Moreover, socioeconomic progress

---

[23] Desai, R.; *Geopolitical Economy: After US Gegemony, Globalization and Empire*, Pluto Press, 2013. *See also*: Desai, R.; *After American Hegemony, Globalization, and Empire*, Bodrunov, S.D., ed.; S. Yu. Witte: Center-Catalogue, 2020.

would be impossible without the concentration of wealth (conditionally investment potential) in the hands of a minority at the expense of the majority. Inequality has, therefore, been an indispensable element in the functioning of economic systems. What has changed is the economic forms that have defined and perpetuated this inequality, using it more or less successfully as a factor in progressive development.

So why has the dream of happiness always been associated with equality? The answer is simple: because "unhappiness" was associated with blatant inequality. The answer is seemingly clear, but not quite right. After all, inequality, as shown above, is an inevitable state of society at a certain stage of development. Moreover, inequality is, to *some extent,* necessary and beneficial for development. The requirement of equality arises when this *measure is breached.*[24]

The demand for equality is thus not a rationally considered program, but merely an expression of protest. "Liberty, Equality, Fraternity" is only a protest slogan addressed to the masses to ignite their dream of a more just society, not a positive program (as a slogan that prominently figures in the French bourgeois revolution, as well as those of the American revolution, and others that were never adopted).

The study of economic inequality (as indeed any economic problem) only makes sense in the period before the formation of a new society as long as the economy exists as a mechanism for satisfying needs. It can offer important insights into the relationship between the level (rate) of satisfaction of needs and the level (rate) of growth in different strata of society at different stages of civilizational development. And it is to be commended that more scholars today are addressing these issues, devoting serious work to them in terms of volume and content.[25]

---

[24] On the social consequences of the growth of inequality *see,* for example, Bodrunov, S. D.; Galbraith, J. K.; *New Industrial Revolution and The Problems of Inequality,* Bodrunov, S.D., ed.; G.V. Plekhanov Russian University of Economics, 2017, 50–51 et al.

[25] There are quite a few studies on this topic. *See,* e.g.: *Inequality of Income and Economic Growth,* Buzgalin, A.; Traub-Mertz, M.; Voeikov, M.; eds.; Cultural Revolution, 2014; Wright, E.O.; Perrone,L.; Marxist Class Categories and Income Inequality; *American Sociological Review* 1977, *Vol. 42, No. 1,* 32– 55; Wolff, E.N.; *Poverty and Income Distribution,* Wiley-Blackwell. 2008; Piketty, T. *Capital in the Twenty-First Century,* Éditions du Seuil. 2013; Stiglitz, J. G.; *The Price of Inequality. How the Stratification of Society Threatens Our Future,* Exmo, 2015; The Global Wage Report 2014/15: Wages and Income Inequality, International Labor Organization; Geneva, 2015. The author's position is outlined in the book: Bodrunov, S. D.; Galbraith, J. K.; *The New Industrial Revolution and the Problems of Inequality,* G.V. Plekhanov Russian University of Economics, 2017.

Such a study gives a glimpse of the proximity of the system (in this case, the socioeconomic system) to rupture, moving toward destruction. Thus, feelings of inequality and injustice are an indicator of tension in society, signaling that the gap between what is possible and desirable on the one hand, and what is actually available on the other, is too wide for most people.

So, if we proceed from the above, inequality will not disappear during the transition to NIS.2 and further to noonomy. It will become different, perhaps just as acutely felt but perceived as inevitable. Its parameters will be carefully monitored to avoid overstressing the public system, and bringing it to a new state in time.

Inequality remains, of course, but it will not be about unequal opportunities to meet one's needs, but about the unequal ability to use and embrace those opportunities open to all. To satisfy one's spiritual needs, one must possess spiritual capabilities. Without a certain level of culture, neither music (even with absolute hearing) nor fiction can be adequately understood. You cannot satisfy your passion for research in a particular field (mathematics, physics, materials science, genetics, etc.) without a certain level of knowledge in that field. In the noo-stage, there will be no *social* barriers to the acquisition of such abilities, and only differences in *individual* abilities will remain as factors of inequality.

Another thing is that the road to such a state of society (and social inequality) is a very, very long one. Therefore, it is necessary to understand the problem of inequality (in its current state and for the future), to identify its sources, possible negative consequences, and ways to overcome it. Then, at the NIS.2 stage, move it from increasing to decreasing in importance for sociodynamics and the state of the socioeconomic system, gradually becoming insignificant. The solution, against the backdrop of increasing opportunities to meet nonsimulative needs in NIS.2, is to recognize the need to limit simulative needs and gradually move toward a noo-type of social consumption, formation, and satisfaction of needs.

The contemporary sociodynamics of inequality calls for an understanding of moving from the current stage through NIS.2 to a noo-type of consumption. When and how do the factors of self-limitation, inner limitation of needs, and refusal of simulated self-satisfaction come into play? In this sense, NIS.2 is a dangerous crossroads: here, a rupture can form when the possibility of unlimited satisfaction of needs is approaching, but the awareness of the need for reasonable self-restraint has not yet been fully achieved.

Growing inequality is an indicator of disadvantage in today's "economic" world. It would seem to push up the entropy of this system. But it is more of an indicator than a baseline cause of a possible explosion. That's not the point. The chaoticization of the system is heightened by an ever-increasing contradiction. The STP provides increasingly recognizable options to meet increasingly recognizable needs, including other types of needs that are gaining importance: access to education, culture, and other intangible knowledge needs. That is on the one hand. And on the other hand, the restriction of access to these options (the consequence of the transfer, transit, transposition of "bio-relationships" to the socioeconomic and sociostructural spheres) is increasing at all levels (between groups of population, regions, countries). Recent tensions, reflected in the emergence of a phenomenon of "new normality" and so on, are created by the accelerating and expanding superposition of technological and social shifts in the space of global civilization.

At the same time, we should not forget that for a large part of humanity, there is still a problem of access to real-life needs: clean water, affordable food, basic literacy, etc. For them, the issue of inequality stands in its primordial, primitive-natural form as a struggle for subsistence. Moreover, according to RAS Academician A. A. Anfinogenova, the projections reflect a "declining average per capita of food production, deteriorating natural resources, and an increasing number of poor countries in need of food aid."[26]

This needs to be borne in mind because it is a problem that raises both enormous conflict potential and the issue of the level of pressure on the Earth's resources.

Further growth of the inequality indicator (and its value can be measured by economosociometric means, as if partially verifying the disharmony of life with this "algebra") leads, despite the general/global movement toward NIS.2, to an exacerbation of conflicts. The lack of consideration of this circumstance is fraught with multiple negative consequences.

---

[26] Anfinogenetova, A. A.; Yakovenko, N. A.; Theoretical and Methodological Problems of Innovative Development of Russian Agro-food Complex; *Regional Economy* 2011, *No.4*, 92.

5.2 REINDUSTRIALIZATION, DIGITIZATION, AND THE NATIONAL TECHNOLOGICAL INITIATIVE

5.2.1 REINDUSTRIALIZATION

The main trend of the new deal is re-industrialization based on the predominant development of high technology[27] and a qualitative *renewal of the technological* basis of material production based on a new understanding of the nature of the global economy. It is about the accelerating nature of change in the economic system, including the main components of the production process mentioned above: its organizational basis; technology, materials, and equipment; the content of labor in production; and, finally, the result of the production process—the product of production.

It is worth underlining once again that the task of "creating a qualitatively new technological base for the industry does not contradict the theses of prominent scientists on the need to move toward a new material basis of production, for example, based on wide application of technologies of the 5th and 6th modes (S. Yu. Glazyev),[28] informatization, miniaturization, individualization, and network organization of production (Castells),[29,30] wide use of the creative potential of workers,[31] and so on. However, it

---

[27] *See*, for example, Bodrunov, S. D.; Grinberg, R. S.; Sorokin, D. E.; Re-industrialisation of the Russian Economy: Imperatives, Potential, Risks; *Economic Revival of Russia* 2013, *No. 1 (35)*, 19–49; Bodrunov S. D. On the Reindustrialisation of The Russian Economy in the WTO Environment; *Economic Revival of Russia* 2012, *No. 3 (33)*, 47–52; Bodrunov, S.D. Reindustrialisation: Round Table at the Free Economic Society of Russia; *World of New Economy* 2014, *No. 1*, 11–26; Tatarkin, A. I.; Sobering Up After the Market Euphoria Is Delayed, But It's Still Happening: Interview; *City 812* 2014, *No. 32*, 21–23; Bodrunov, S. D.; Grinberg, R. S.; What To Do? Imperatives, Opportunities and Problems of Reindustrialisation; Mat. Scientific and Expert Council under the Chairman of the Federation Council of the Russian Federation: Reindustrialisation: Opportunities and Limitations; *Council of the Russian Federation*, 2013; Bodrunov, S. D.; Reindustrialisation of Russian Economy Opportunities and Limitations; *Proceedings of the Free Economic Society of Russia* 2014, *No. 1*, 15–46.

[28] Glazyev, S.Y. *On External and Internal Threats to Russia's Economic Security in the Context of American Aggression: a Scientific Report*, 2014.

[29] Castells, M.; *Information Epoch: Economy, Societies and Culture*, Shkaratan, O.I., ed.; Higher School of Econ., 2000.

[30] Bodrunov, S. D.; *Coming: The New Industrial Society: Reengineering*. Edition 2, INID, S. Yu. Witte, 2016.

[31] Buzgalin, A. V.; Kolganov, A. I.; Reindustrialisation as Nostalgia? Polemical Notes on The Target Emphases of Alternative Socioeconomic Strategy; *Sotsis* 2014, *No. 3*; Krasylshchikov, V. A.; Modernization and Russia on the Threshold of the XXI Century; Voices of Philosophy 1993, No. 7, 54–55; Sakaya, T.; Value Created by Knowledge, or History of the Future; *New Post-industrial Wave in the West: An Anthology*, Kovalev, V.V.; Inozemtseva, L., eds., 1999.

opposes the ideas of supporters of vulgarized "post-industrialism," who advocate the *priority* development of the sphere of non-productive services, intermediation, and financial transactions. Misunderstandings are the result of methodologies in scientific analysis which are based on viewing the economic system either in a "photographic-static" state or in a certain dynamic, but almost always without taking into account "dynamics of dynamics," acceleration and second derivatives, which continuously and with different rates of acceleration change the nature of the phenomena, processes, system elements, their relationships, etc., to be analyzed.

Post-industrial concepts were based on the real processes of transnational corporations transferring some industrial production capacities from developed to developing countries. However, as A. A. Shirov notes, "Any process has limits. The declining share of basic industry in GDP formation has a negative impact on both employment and income. The share of revenues held by multinational corporations is becoming unacceptable even for developed countries. In this context, the modern reindustrialization of developed countries has become one of the key elements of the modern economic agenda."[32]

The main goal of reindustrialization as *an economic strategy* is *restoring the role and place of industry* in the country's economy in the process of its *structural adjustment as a basic component* and *priority development of material production* and the real sector of the economy based on *a new, advanced technological pattern* within the framework *of Russia's modernization.*[33]

One of the *consequences* of deindustrialization is reduced efficiency of Russia's integration in the global division of labor. This is reflected by the presence of Russia only at the initial stages of value chain creation in the most basic sectors, which creates a *technological dependence* on developed countries. Russia specializes in extraction, production, and supply to international markets of low value-added products: natural gas, oil, ferrous and non-ferrous metals, potash fertilizers, etc. Russia's high-tech exports are mainly related to weapons and military technology (there are also nuclear industry, space technologies, and production of

---

[32] Shirov, A. A.; Socioeconomic Forecast as a Mechanism of Strategic Economic Management; *Budget* (Online) 2019, *V. 1.* http://bujet.ru/article/364772.php

[33] Bodrunov, S. D. *Formation of Russia's Reindustrialisation Strategy.* Edition 2. In two parts: Part One. INID, 2015. Part Two. INID, 2015.

titanium products, but against the general background, their volumes are relatively small).

"The analysis and international practice show that successful *reindustrialization* (including export-oriented *import substitution*) requires at least two economic policy priorities:

The first is *a favorable economic environment* including availability of resources, reduction of administrative barriers and bureaucratic pressure, tax incentives for industrial enterprises, their preferential long-term lending, increased protection of investments and assets (rights and property of investors), etc.

The second one is *an active state industrial policy* aimed at priority development of key spheres of material production (primarily science-intensive high-tech), as well as science and education.

Active industrial policy in a broad sense implies:

- *the adequate monetary policy of the Central Bank and the fiscal policy of the Ministry of Finance*, which would ensure financing the development of industrial and agro-industrial complex enterprises in the required volume;
- *the stimulation of domestic demand* for the products of industrial enterprises, including through "pre-tax" prices and the system of state order;
- *the long-term nature of these activities*, which allows for attracting long-term investments;
- *maintaining a high degree of openness in the economy* (except for industries that ensure the economic capacity and security of citizens); development of cooperation with foreign partners technology exchange, scientific cooperation, and creation of advanced production technologies;
- and *state support for exports* of competitive industrial products.

This leads to a *fundamentally important conclusion*: proactive industrial policy, public–private partnerships, selective protectionism and international cooperation in production, science, and education are all needed to address these challenges.

It should also be noted that a proactive industrial policy is an indispensable tool when we are lagging behind technological and industrial leaders. According to V. E. Dementiev, "Selective industrial policy bears the main burden of counteracting the trend where technologically leading countries remain points of attraction for investment resources even during the global financial crisis."[34]

Risks should not be forgotten, of which the most important are:

1. *lower competitiveness of Russian industrial products* due to "sterile conditions" for the development of Russian industrial enterprises (availability of state support and lack in the domestic market to compete with leading foreign manufacturers). The result is a reduction in management quality and an increase in the quality and price of the products. The decision to use foreign components to develop the Sukhoi Super-Jet 100 relates to the latter circumstance, as Russian suppliers could not provide a competitive price-quality ratio. It is clear that the main way to address this problem is through the development of domestic innovation,[35] targeted applied research, and industrialization of the results, which require closer integration of science and industry;

2. *reduction in the efficiency of the country's economy as a whole* if the products and technologies of domestic producers are inferior in terms of competitiveness (price, quality, range) to their foreign counterparts. This situation is characteristic of the development and production of oil and gas equipment in non-standard geological and natural climatic conditions. In addition, for example, a deterioration in the quality of domestic medical equipment or medicines can lead to a significant reduction in the quality of life of the population. Thus, an import substitution policy in industry without a systematic approach to its implementation (including continuous monitoring of the dynamics of industrial development in terms of industries and enterprises) may lead to a decrease in the competitiveness of the national economy as a whole. This is a systemic risk, which is determined by the inefficiency of the institutional environment;

3. *increase the burden on the budget*. The implementation of import substitution policy as part of the reindustrialization strategy requires significant investments from the government. For example, the federal target program for developing the defense

---

[34] Dementiev, V. E.; *Long Waves of Economic Development and Financial Bubbles*, CEMI RAS, 2009.

[35] Tsatsulin, A. N.; Approaches to Economic analysis of Complex Innovation Activity; *Proceedings of St. Petersburg State University of Economics* 2013, No. 2 (80), 12–21.

and industrial complex for 2011–2020 was allocated 3 trillion roubles by the government. If the economic situation deteriorates (which is the case in modern Russia) and it is impossible to meet planned budget expenditures, the government must either reduce spending on the social sphere and other areas or suspend funding for import substitution activities. As a result, the risk of growing *corruption increases*. The representatives of state corporations and officials have the opportunity (and temptation) to lobby for decisions related to the reallocation of scarce budgetary resources;

4. *technological lagging of the Russian industry at the global level*, caused by two circumstances. First, with a long import substitution process there is a risk of the partial substitution of imports from economically developed countries to affordable imports from Asia, Latin America, and Eurasian Economic Union (EAEU) partners. This will not only retard the development of the industry but also consolidate the trend of lagging behind the technological level of Russian industry (which today is 40–60 years). Secondly, the strategy of import substitution in the short term focuses on replacing foreign products with domestic analogues. In essence, we are talking about copying foreign products and technologies that exist on the market—and that means a permanent technological gap. This risk can be overcome by developing the domestic research, design and technology base, and schools ahead of production, which requires greater efforts to support science and education and integrate them with production.[36,37]

It should be noted that political will alone, even if backed by financial resources, is not enough to implement import-substituting reindustrialization. The complex and ambitious task of rebuilding high-tech material production requires a strategy oriented toward science, world-class education, and an advanced level of culture, which will continue to make Russia proud and which citizens of other countries will aspire to.[38]

---

[36] *See:* Krasilshchikov, V. A. *Follow-up to the Past Century: Russia's Development in the Twentieth Century from the Perspective of World Modernizations*, ROS-SPEN, 1998; Russian State Library of Russia, 2010.

[37] Integration of Production, Science and Education as the Basis for R-Industrialisation of the Russian Federation; *World Economy and International Relations* 2015, No. 10, 100–102.

[38] Bodrunov, S. D.; Integration of Science and Education Production and New Industrialisation of Russia; *Vedomosti, No. 215*, 17.

Re-industrialization of this type is possible only in *a modernized institutional environment*. Most experts believe that a poor institutional environment is the main constraint to economic growth in Russia. The effect of institutional changes is comparable to, or may even exceed, the effects of fiscal and monetary stimulus measures.

Economic modernization calls not so much for the development of competition in general but rather for creating conditions where Russian entrepreneurs would be *forced* to use technological modernization as the main tool of competition. The abolition of monopoly to promote competition is also *necessary but not sufficient.* It is key to achieve a change in the nature of the appropriation of the result of economic activity. If redistribution of property rights is much more attractive than development, the fight against corporate raiding and calls for innovative behavior will continue unabated. Compensation measures (credit-investment-tax incentives or mechanisms of public-private co-financing) are not sufficiently significant factors of innovative activity risk reduction. Institutions that make the use of other (non-innovative) competition tools significantly more risky are much more effective.

5.2.2 DIGITALIZATION

The most important element of the infrastructure of the modern economy is its information support. Digitalization of assets and economic management structure is one of the most urgent directions of reindustrialization and the creation of advanced technological industries. Undigitalized assets are losing market value and obsolescence is accelerating under the influence of the global trend toward info-digitalization, which does not "fit" into current usage patterns. Ownership of such assets, even the most advanced ones, does not improve the competitiveness of the economy, but on the contrary, requires excessive funds to maintain them.

The industrial complexes of industries have the potential for economic growth and require digitalization as a priority.

Digitization of assets should be carried out at all levels: enterprise (finished products, business processes such as warehouse production realizations, management systems, etc.), cooperative group, and industry. The digitalization of the industry enables the construction of cross-industry platforms that quickly increase the efficiency of cooperation groups, reduce transaction costs, reduce unnecessary elements of transactional chains, intermediaries, etc.

Evidently, the economic infrastructure customs, transport and logistics, roads, fiscal, etc. should also be digitized, allowing for a radical increase in the efficiency of freight, goods, and services exchange. In addition, a "digitized" participant economy, and only that, will be able to ensure that its actors are included in the most advanced segments of the global marketplace of the coming decades.

However, as already noted, digitalization on its own, without the support of sixth-stage technologies, for which it is a means of integration, will not have much effect. It is not feasible to make technological breakthroughs without a state-of-the-art industrial base and a policy of reindustrialization.

5.2.3 ECONOMIC AND INSTITUTIONAL CONDITIONS FOR MODERNIZATION

The global trend of the global economy in the twenty-first century shows not growth, but the decline in demand for traditional materials, raw materials, and energy. This is inevitable when the role of industrial knowledge, technology, its acquisition, absorption, implementation in the real sector, development, etc., increases dramatically. The decline in oil and gas prices, which has been underway for several years now, is a harbinger of a new era: natural resources will be much less important for developing a new industrial economy in the NIS.2. The transition to a so-called low-carbon economy, characterized by a reduction of the environmental load by reducing the use of fossil fuels and CO_2 emissions, has been a global challenge for years. It is not just a question of the technological challenges, or the costs involved, but of reconciling the transition to a low-carbon economy with progress toward meeting our society's social and economic development challenges.[39] This is not possible without a broader reliance on the application of new scientific knowledge.

This very change in the ratio of material to knowledge in the final product allows us to expect that future generations will be able to utilize unspoilt natural resources. But for this to happen, developed countries (including Russia) must constantly own advanced technologies and use them intelligently.

---

[39] *See*, for example, Porfiriev B.N.; Low-Carbon Development Paradigm and The Strategy Of Climate Change Risk Reduction For The Economy; *Problems of Forecasting* 2019, *No. 2*, 3–13.

Technological breakthroughs can only be achieved with precisely targeted programs that implement long-term strategies. But they, too, can only be successfully implemented if the participants' interest in such programs is secured.

The effectiveness of such an implementation mechanism requires the adoption of relevant laws and regulations. Once again, the role of the national innovation system (NIS), the most important mechanism in modern conditions for shaping and implementing modernization objectives, is great in any country. Russia requires developing a new methodology for assessing efficiency and effectiveness of NIS and its components, based on a systematic approach, with a focus on an intensification of innovation renewal in Russian industry, creating prerequisites for achieving *technological leadership* in the world in *selected areas*; and developing effective mechanisms for the *transformation of innovation potential* into *new technologies* that are in demand on the market.

Another serious reason for the insufficient performance of the Russian NIS is *the lack of a developed market of innovative products, services, and technologies.* The underdevelopment of *the intellectual property market poses a major challenge.* The analysis of the content of more than 150 federal, regional, and sectoral strategies and programs of innovative development by industry, carried out by the S.Y. Witte Institute of Scientific Research in cooperation with the Republican Research Institute of Intellectual Property (RRIPI), revealed that the development of the intellectual property market—one of the "driving belts" of innovative development—is not considered at all in most of these documents.

The State Programme for the Development of Science and Technology,[40] adopted in Russia as a founding document in this area, was "designed to create a competitive research and development sector in Russia, capable of ensuring the technological modernization of our economy."[41] It was planned that from 2013 to 2020, the financing under this program would amount to 1 trillion 187 billion rubles. The program's main objectives

---

[40] The State Program "Development of Science and Technology" for 2013–2020 (Approved by Government Decree of April 15, 2014

No.301); Website of the Ministry of Education and Science of the Russian Federation. http://минобрнауки.рф/%D0%B4%D0%BE%D0%BA%D1%83% D0%BC%D0%B5%D0%BD%D1%82%D1%8B/2966 preliminary stopped by the Government Decree No. 377 of 29 March 2019).

[41] Bodrunov, S. D.; Innovative Development of Industry as a Basis for Technological Leadership and National Security of Russia; *Proceedings of the Free Economic Society of Russia* 2015, *No. 3, V. 192*, 44.

were to finance scientific and technological advances for further use in the programs of the relevant authorities and to support interdisciplinary research, which has been associated with promising advances in science and technology in recent decades.

In the process of realizing these objectives, expenditure on funding scientific activities has grown steadily. The share of the federal budget expenditure on science is gradually increasing, as well as their share in GDP, remaining, nevertheless, at a low level of 0.5–0.6% (Fig. 5.2).

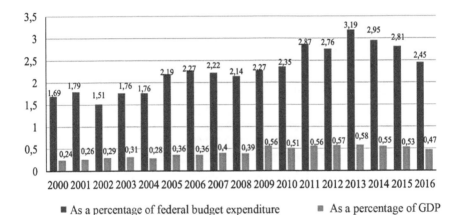

FIGURE 5.2 Financing of science from the federal budget, share in GDP.[42]

However, measures taken by the state to develop the innovation system and fund research are clearly insufficient. The number of organizations engaged in scientific research has fallen significantly (from 4,099 in 2000 to 3,605 in 2013), while the number of economic entities and GDP has grown. The attractiveness of research activities for companies and organizations is decreasing (Fig. 5.3). According to Rosstat, the main developments are increasingly carried out by specialized research organizations, which are not sufficiently close to production processes. This demonstrates

---

[42] 2000–2004: Russia in Figures, *Federal State Statistics Service*, 2007. http://www.gks.ru/bgd/regl/b07_11/IssWWW.exe/Stg/d020/21-07.htm; 2005–2009: Russia in Figures, *Federal State Statistics Service*, 2011. http://www.gks.ru/bgd/regl/b11_11/IssWWW.exe/Stg/d2/22–07.htm; 2009–2013; Russia in Figures, *Federal State Statistics Service*, 2014. http://www.gks.ru/bgd/regl/b14_11/IssWWW.exe/Stg/d02/22–07.htm; 2014–2016 Russia in Figures, Federal State Statistics Service, 2018. http://www.gks.ru/bgd/regl/b18_11/IssWWW.exe/Stg/d02/22–08.doc)

the continuing disintegration of science and industry, which slows down innovation and new technology.

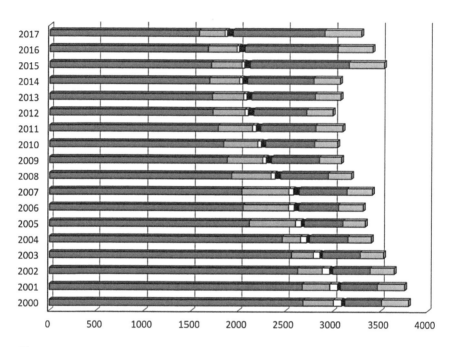

■Research organizations ■ Design bureaus □ Design and survey organizations
■ Pilot plants ■ Higher education institutions ■ Industrial organizations with research, design, and engineering divisions

FIGURE 5.3 Structure of participants in scientific activities.[43]

5.2.4 NATIONAL TECHNOLOGY INITIATIVE

Strengthening Russia's role in the global division of labor requires developing, regulatory consolidation, and implementing a systematic, integrative document.

First of all, we are talking about long-term documents of strategic importance, such as the "Conceptual Framework for the National

---

[43] Number of Organizations that Performed R&D by Type of Organization in the Russian Federation, *Website of the Federal State Statistics Service*, Updated 28 August 2018. http://www.gks.ru/free_doc/new_site/business/nauka/t_1.xls

Technology Initiative (NTI)."[44] The Russian Academy of Sciences developed this document at the request of the President of Russia, based on his Address to the Federal Assembly of the Russian Federation on December 4, 2014. The Russian Venture Corporation (RVC) provided 750 experts to develop the NTI roadmaps. The RAS Working Group also worked out a document, according to which the country's technological leaders of the next 10 years should comply with several requirements:

- having a clear and coherent science, technology and *innovation policy* that *focuses on technological leadership* and is supported by the *necessary resources*;
- *diversity of forms of scientific research organization;*
- *a knowledge-intensive industry* based on proprietary technology;
- *education* focused on the training of creators;
- *business as the main investor* in research and development;
- business working for *the development of society.*

Russia needs to solve several problems to join the ranks of technological leaders:

- modernizing production facilities;
- supporting and stimulating the innovative activity of enterprises;
- financing research and development activities, and developing new technologies;
- training highly qualified personnel of different levels—workers, scientists, teachers, and managers;
- actively developing an innovative infrastructure designed to help bring science and business together.

The goals stated in this document can only be achieved when the technological levels of the economies of Russia and the leading countries are comparable.

This ambitious program to accelerate the country's economic development by supporting high-tech start-ups and NTI companies was launched in 2016. In 2017, the Government of the Russian Federation approved road maps for the development of technological renewal in several sectors of the economy.

---

[44] Website of the Russian Academy of Sciences. http://ras.ru/viewnumbereddoc aspx?id=69fa7c74-4033-4215-b908-911a87acf803&_Language=ru.

It was clear that Russian companies would not be able to squeeze competitors from developed countries out of their high-tech markets. NTI participants were tasked with identifying the most current trends in technology development; catching the wave of the new technological revolution; and charting the entry of domestic entrepreneurs into high-tech markets that are at the very beginning of the birth process. Thus, NTI is aimed at implementing the concept of accelerated development: "overtaking without catching up."

Advanced technological solutions must meet three types of challenges: developing new technology sectors, ensuring national security, and improving the quality of life of Russia's population.

NTI focuses on the development and application of technologies for which Russia has scientific and economic potential. It is equally important to identify the export potential of promising high-tech industries. The point is that the size of the Russian domestic market is not sufficient to build production that can meet the challenges of global competition. The export orientation sets the necessary bar of requirements: to do better than others to conquer the global market.

To this end, support mechanisms had been developed during the initial phase of NTI development:

- building NTI infrastructure;
- searching for and training talented young people;
- direct financial support for the creation of technology;
- support for innovation-active companies through tax incentives;
- assistance in exports and promotion of NTI companies in foreign markets.

In 2019, about 20% of the projects supported by NTI had already started selling. By the end of 2019, 150 projects on end-to-end technologies were launched: artificial intelligence, virtual reality, quanta, sensorics and robotics, and big data storage and analysis.

At the same time, proposals for legislative initiatives reducing barriers to high-tech business development were prepared. Out of 60 proposed draft laws and regulations, 40 have already been approved. On 3 November 2018, the Government of the Russian Federation signed Order No. 2 400-r on the establishment of the Autonomous Non-Profit Organization known as the "National Technology Initiative Platform." The NTI platform combines data from Russian (and, in the future, international) development institutions, foundations, accelerators, and the Leader ID system, as well as information on conferences and competitions into an integrated system. This data allows the platform to "follow" projects as they develop—identifying the stage young companies are at, suggesting appropriate support measures, and offering expertise.

At the beginning of the third decade of the twenty-first century, NTI is entering a new phase by implementing the NTI 2.0 project. Advancements in technology are driving the formation of new markets. That's why NTI 2.0 aims to bring entrepreneurs together around new growth points. The ability to see new markets gives an understanding of where the demand will be, what people will want to consume in the future, and where "the money will come." This makes it clear where the ground will be for new start-ups.

The aim is to move to a fully digital model of NTI system interaction. Currently, the processes that underpin NTI are not fully digitalized, and there is a lot of potential for development here. The challenges that hit the global economy in 2020 have shown that online business technologies have a higher sustainability potential than offline ones.

In addition, digital collaboration technologies create a more conducive environment for engaging regional actors in NTI processes. However, by the end of 2020, NTI has the status of a government program only. It is imperative to elevate this status to a national project, as all of the approved national projects have technology development targets. However, in order not to make a mistake in choosing directions, a ready-made NTI tool should be applied.

The National Technology Initiative is a science-based blueprint for our technological development, developed and implemented by a large group of highly competent specialists. It includes both areas of development and basic ideas, and a comprehensive program of measures to create fundamentally new markets and conditions for global technological leadership in Russia by 2035. At the same time, the initiative is developing under the influence of global technological development trends. It is a carefully prepared, expert-validated, interdisciplinary program with a well-established governance structure (which is crucially important!). This is one of the reasons why it is advisable to give NTI governance the status of a national project.

National projects are characterized by specific, measurable indicators and the personal responsibility of project managers, which facilitates monitoring of their implementation. This is a more efficient way of reaching goals than working through state programs, where ministerial coordination leaves a lot to be desired.

Moreover, national projects have more stable and aligned objectives. This is crucial for implementing the state development strategy. For business investment planning, this is also important as the basic plans of the state should not change with every new budget law. Obviously, businesses are more likely to invest within a specific goal designated as a government priority than simply at their own risk.

Innovations mustn't emerge on their own but as a response to demand and to help meet customer needs. The state sees technological progress as a means of solving social, medical, and scientific problems. Therefore, the most promising development of the Russian economy is the integration of the NTI into national projects, ultimately aimed at improving the quality of people's lives. In this case, technological development is justified and demanded, and money for creating innovation is not wasted (or spent on interesting but economically inefficient issues). This is where both the objectives of specific national projects and the problem of realizing these objectives at an advanced level of technology as well as the main challenge of the technological modernization of the entire Russian economy are being addressed simultaneously.

However, the objective of Russia's economic modernization is to shift from a technological multi-economy dominated by the 3rd and 4th orders to a higher level of the 5th and 6th orders. Therefore, according to a group of leading Russian scientists, the development strategy urgently requires "consideration of the resource-technological heterogeneity of the Russian economy, the presence in its structure of backward industries, [and its] functioning on the basis of low quality and low efficiency resources and technologies, which cannot be overcome when applying universal tools of market economy regulation and require the development and implementation of a complex of specialized individual measures."[45]

However, social safeguards must not be forgotten in the new economy, when innovation will begin to crowd people out of the production process. The social program should work in tandem with the modernization program.

---

[45] Govtvan, O. D.; Gusev, M.S.; Ivanter, V. V.; Xenophontov, M. Y.; Kuvalin, D. B.; Moiseev, A. K.; Porfiriev, B.N.; Semikashev, V. V.; Uzyakov, M. N.; Shirov, A. A.; System of Measures to Restore Economic Growth in Russia; *Problems of Forecasting* 2018, *No. 1*, 4.

At the beginning of this decade, the task was set to revive Russia's industrial complex. In 2020, due to the global crisis caused by the coronavirus pandemic, the global economic environment was not favorable for an abrupt start to the reform program in the Russian economy. However, this task cannot be dismissed all the more so because the economic situation is unfavorable not only for Russia, but also for its global competitors.

5.2.5 INTEGRATION OF PRODUCTION, SCIENCE, AND EDUCATION

Integration of production, science, and education is one of the principal positions of state economic regulation in several leading industrialized countries. In Japan, for example, cooperation between industry, science, and government has been a strategic direction of the government's innovation policy for many years. Since the mid-1990s, Japan has enacted a series of laws that have fostered and strengthened ties between the private sector, science, and government. In 1995, the Law on Science and Technology came into force, providing for state financial support for research at universities. In 1998, the Technology Licensing Organization Promotion Law was introduced, allowing companies to benefit from universities' research and development activities through specially created organizations. The Act on the Support for Production Technology Development (2000) has enabled public university professors to set up their own private companies to ensure that the results of their research are applied in the industry. Moreover, one of the main objectives set for universities was to support the development of production technologies. Finally, a major law on intellectual property was adopted in 2002, defining a framework for cooperation between industry, science, and the state to stimulate the development of the country's economy by using the results of scientific research activities.[46]

In accordance with this law, Japan is actively implementing programs for the development of science and technology cooperation between innovation process participants. Japan has been driven to implement such programs by the U.S., which has significantly increased its competitiveness in biotechnology and information and communications technology (ICT) through such programs.

---

[46] Bodrunov, S. D.; Integration of Production, Science and Education as the Basis for Reindustrialization of the Russian Federation; *World Economy and International Relations* 2015, *No. 10*, 97.

Another telling example is Germany. "Some of the main initiatives and projects of the German government in this area are:

- integrating science, education, and industry, and the state support of innovation clusters with the participation of small and medium-sized enterprises and scientific organizations (projects of the Association of Industrial Unions named after Otto von Guericke);
- implementing targeted innovation projects in the new federal states;
- developing new instruments for financing promising innovation clusters;
- organizing a federal competition [titled] "Germany's best innovation cluster" with the participation of universities and colleges;
- improving public–private partnership models in the development of innovative activities;
- further improving the system of scientific personnel training and their involvement in research activities.

Let us add that the integration of production, science, and education is a powerful trend in the development of modern global industry: the development and implementation of various projects aimed at establishing and strengthening the system of technological cooperation between business and science in the United States and industrialized European countries started as early as the mid-1980s and early 1990s."[47] In the United States, the famous Bayh-Dowell Act and other legislation have played an influential role.

The reviewed domestic and foreign experience should be critically implemented during import substitution policies as part of a reindustrialization strategy.

Note that in Russia over the last 10–15 years, there have been positive trends in the integration of production, science, and education. The work of FSUE Khrunichev State Space Research and Production Centre, Aerospace Equipment Group, and others can serve as examples of successful integration projects implemented in the first decade of the twentieth century. Following the strategy of space-rocket industrial development, as well as the FTP "Reforming and Developing the Defense Industry Complex (2002–2006)," approved by the RF Government Decree No. 713 on October 11, 2001, a large integrated structure for the development and production of heavy launch vehicles was formed based on Khrunichev

---

[47] *Ibid.*, 98.

State Research and Production Space Center. The most pressing task of integration is to maintain the company's industrial, scientific, and technical potential and to ensure the fulfilment of state orders. The Khrunichev Complex has initiated integration with several leading Russian technical universities, providing the targeted recruitment of students to work in the complex's enterprises and design bureaus.

Similar microlevel projects are known in creating innovation clusters, building technology transfer networks,[48] technology hubs, etc.

However, Russia still lacks a long-term working strategy for integrating production, science, and education at the macro level. Tasks are mainly handled in "manual mode."

Based on lessons from domestic and international experience, as well as a summary of theoretical perspectives, *recommendations for measures to reintegrate production, science, and education can be formulated.*

Firstly, the material and technical basis for innovation in the production-science-education integration project (PIE) should be based on the solution of well-known problems:

- the education system is designed to train creative people, specialists, and professionals;
- the deployment of research and development based on the achievements of basic science;
- bringing new technologies to industrial designs;
- and the organization of mass serial production of such products at domestic enterprises.

However, in the current situation, these requirements can initially only be fulfilled in limited areas.

Therefore, *secondly,* modern Russia should focus primarily on the revival of *the preserved reserves of high-tech modes* (mainly in the defense sector), and programs of complex creation of new technologies and fundamentally new products to be implemented in a limited scope and only in the directions that promise the greatest economic effect for the national economy.

Thirdly, the economic mechanisms for this project may be based on market-based instruments (financing through public contracts, long-term loans, and guarantees), public–private partnerships, long-term government

---

[48] Osipenko, A. S.; Technological Transfer in The System of Innovation Development of Industry; *Economic Revival of Russia* 2014, *No. 1 (39),* 83–88.

programs and active industrial policy, linking market-based mechanisms with public investment, and development plans for state-owned enterprises (including education and science).

Fourthly, institutional support for these priorities may include special institutions of long-term development (ensuring the development and implementation of strategic programs, active industrial and structural policy, etc.). To be successful, they need a reduction of administrative barriers in the financial, credit, tax and customs systems, and more state support in the areas of patenting, certification of technological processes and products, etc.

Integrated production, science, and education clusters of different organizational and legal forms from open networks to complexes that have a common development program and work for a common long-term result, with unified financing and coordinated management, can play an important role here. The application of one or another form depends on the specific tasks to be performed and the existing prerequisites.

CHAPTER 6

Strategizing on the National, Regional, and Sectoral Economies

6.1 DEVELOPMENT OF OPPORTUNITIES FOR STRATEGIZING THE TRANSITION OF THE NATIONAL ECONOMY TO THE NIS.2

6.1.1 PLANNING AS A NECESSARY TOOL FOR IMPLEMENTING STRATEGIC PROJECTS

As shown in the previous sections, *the strategic plan is the fundamental element in the strategy implementation, and the main management tool for its execution*. Without concrete plans and programs, the strategy is not feasible.

In general, planning should be evaluated as a higher-order phenomenon than chaos in terms of entropy reduction and ordering the dynamics of system development. Humans have always struggled with the chaos and uncertainty of existence. In this sense, planning is the next step compared to non-planning, for example, with the market in achieving a higher level of sustainability in the socioeconomic system. The development of civilization moves toward increasing the elements of planning in economic development.

The experience of China, which has not abandoned planning as an institution and an instrument of development management, indicates that both the economy and society are moving toward a new type of noo-industrial society using such tools. It is hardly possible to imagine a future society, an intellectual noo-society, without the institution of planning as one of the main, basic instruments of social management, and of its entire being.

"We should not be afraid of the word 'planning'," emphasizes B.N. Kuzyk. Today, we do not know an effectively developing country that has not engaged in planning to implement its strategic objectives. In Russia, therefore, a completely new system of long-term forecasting and strategic planning must be developed, approved, and implemented, with sound legislation, an effective budgeting system, and solutions to the critical human resource problem that exists not only in science, but in all areas of activity, to create a new economy. The federal, regional, and municipal levels must be in harmony with each other since it would be impossible to build or implement a long-term strategy without similar work on long-term visioning and strategic planning in the regions. Finally, it is critical that an innovative partnership between science, education, government and business involving civil society takes place."[1]

Without this tool, the digitalization of the economy cannot be implemented in a planned manner.

At least an active industrial strategy and strategic planning in a market economy is *needed for the current level of technology, where the industry is dominated by the 4th and 5th technological modes*. This conclusion overlaps with the ideas of John Kenneth Galbraith.[2]

Unsustainable economic applications of technology can result in new, innovative technologies becoming a Solow paradox[3] rather than development when the introduction of the new does not accelerate development but hinders it.

We must understand how a proactive industry strategy functions. It is like working in the economy of a kind of travelator when each business chooses the most efficient one for its conditions. It is a matter of selective, indicative, or indicative-selective planning. Without these tools, it is unlikely that the social technologies defining the transition to NIS.2, and from it to noo-production, can be used effectively (Fig. 6.1).

---

[1] Kuzyk, B. N;. How To Successfully Implement the Strategy of Innovation Development Of Russia; *Mir Rossii* 2009, *No. 4*, 17.

[2] *See:* Bodrunov, S. D;. *New Industrial Society of the Second Generation: Rethinking Galbraith*, Galbraith: The Return, Bodrunov, S.D., ed.; Cultural Revolution, 2017.

[3] *The Solow paradox* is based on the conclusion of Nobel laureate R. Solow (1987) that the introduction of computers does not lead to productivity growth. Since then, there have been many studies both confirming and refuting this conclusion. It is safe to say with some certainty that this paradox is due, firstly, to the fact that the benefits of information technology require a long period to accumulate the "critical mass" of implementation and, secondly, to imperfect methods for measuring the effects of new technologies, including attempts to measure them only by GDP dynamics. For more details *see*: Platonov, V. V.; "Solow's Paradox" Twenty Years Later, or On the Study of the Impact of Innovation in Information Technology on Productivity Growth; *Finance and Business* 2007, *No. 3*, 28–38.

Market is based on independent decisions of autonomous subjects, based on a spontaneously formed supply demand balance. This requires manufacturers to respond quickly and efficiently to demand. At the same time, autonomous decisions by actors with divergent interests undermine the coherence of the reproduction process and lead to periodic crises, while price-based decision-making criteria narrow the horizon for rational choice of production and consumption patterns.

Directive planning is based on binding decisions for economic entities developed by national planning authorities. It ensures a high degree of coherence in the reproductive process and allows for large-scale reallocation of resources and their concentration on critical production goals. At the same time, it is marked by a delayed response to changes in the structure of needs and is characterized by attenuation and distortion of information signals in both "bottom-up" and "top-bottom" movements. The trend toward a gradual increase in centralization and the extension of directive coverage of various aspects of economic activity is taking on an unsustainable scale.

Indicative-selective planning is based on adopting only the most important production targets and approving binding indicators not for the economic entities but for the planning authorities. The economic entities are oriented to achieve the plan's targets by applying a set of economic incentives.

FIGURE 6.1 Ways of coordinating economic activities in social production.

6.1.2 PROBLEMS AND PROSPECTS OF STRATEGIC PLANNING IN RUSSIA

Practice proves that excluding strategic planning from the main instruments of public administration leads to prevailing inertia in development, an inability to overcome the "rut effect" (which prevents correction of deep structural distortions), accumulation of systemic risks, and does not allow for the achievement of any ambitious goals.

After a long period of disregard for the use of planning tools, including for the implementation of strategic projects, Russia adopted the Federal Law on Strategic Planning in the Russian Federation[4] (hereinafter Law No. 172-FZ) on June 28, 2014. The law regulates the preparation and approval of documents related to implementing the strategic development goals at the federal level.

The drafting of documents defining the country's socioeconomic development strategy took place both before and after adopting this law.

---

[4] On Strategic Planning in the Russian Federation: Federal Law of the Russian Federation of 28 June 2014, No. 172-FZ; *Rossiyskaya Gazeta* 2014. http://www.rg.ru/2014/07/03/strategia-dok.html.

In recent years, for example, the National Security Strategy of the Russian Federation, the Economic Security Strategy of the Russian Federation until 2030, the Strategy for Scientific and Technological Development of the Russian Federation, and the State Policy Framework for Regional Development of the Russian Federation until 2025 have been developed and approved by Presidential Decrees. The Government has approved the Strategy of Spatial Development of the Russian Federation and the Strategy of Innovative Development of the Russian Federation for the period up to 2020 (developed before the adoption of the Law No. 172-FZ). Presidential Decree No. 204, titled "On the National Goals and Strategic Development Objectives of the Russian Federation for the period until 2024," of May 7, 2018 was approved. However, all efforts to develop strategic documents have not resulted in a single strategic plan as a realistic vehicle for implementing any strategic vision.

The limitations of Law No. 172-FZ do not provide sufficient prerequisites for making strategic planning an integral element of the system of management of socioeconomic development of the country. There is a bias in the law toward formal, bureaucratic regulation of the strategic planning documentation process. It does not define the relationship between strategic planning and the current state management tools for socioeconomic development.

Article 7(7) of the Law provides that "participants of strategic planning are responsible for the timeliness and quality of development and the adjustment of strategic planning documents, implementation of measures to achieve the goals of socioeconomic development and national security of the Russian Federation, and for the effectiveness and efficiency of addressing the tasks of socioeconomic development and ensuring national security of the Russian Federation."[5] However, no mechanisms have yet been put in place to hold officials accountable for achieving specific goals of the strategic plans according to specific criteria. And if such mechanisms do not work, the responsibility of officials for implementing strategic plans will remain purely declarative or will become a field of subjective evaluations.

The recently developed proposals for improving public administration suggest separating strategic planning into a specific institutional block

---

[5] *Ibid.*

of the governance system and ensuring the interaction of this block with institutions of long-term, medium-term, and current management.[6]

Strategic planning cannot be understood as a synonym of medium-term planning, which is also oriented toward achieving certain goals in the future. But these goals are constructed as a continuation and development of existing processes and trends, including, of course, their adjustment. Visioning the future in strategic planning involves achieving goals and priorities that are not explicitly embedded in the present, and achieving them involves a qualitative transformation of reality. Otherwise, such goals cannot be classified as strategic.

Strategic planning builds on previously developed and approved strategies that include strategic priorities and objectives. To realize long-term priorities and goals, the strategy process should be based on a broad scientific expertise, public opinion analysis and a business perspective. They are expected to be involved in the development of the strategy and its resulting programs, projects, and plans.

This kind of strategic planning developed in Japan after World War II. The completion of Japan's economic recovery has made it possible to
set strategic goals such as the achievement by key industries (automotive, shipbuilding, and electronics) of the technological level and quality that will ensure global competitiveness. Then a new strategic goal was set, titled the "intellectualization of production," which implied a bet on the development of knowledge-intensive industries. Active industrial policy and economic programming were the tools used to implement the strategic priorities.

The distinction between strategic planning and monitoring requires that these subsystems are closely linked. It requires continuous monitoring of the implementation of planning documents, monitoring the achievement of the objectives set, and making the necessary adjustments to the objectives set according to changing circumstances.

However, in Russia, evidence-based strategic planning methodology has not yet been disseminated at the national level. Strategic planning documents are often developed as various programs and plans with only targets, and without elaborating the resources and mechanisms for "transforming" the present into the desired future. Considering the overall situation, one can say that Russia lacks a national development strategy.

---

[6] Smotritskaya, I. I.; The New Economic Strategy Requires a New Quality of Public Administration; *Bulletin of the Institute of Economics of RAS* 2017, *No. 5*, 15–17.

The consequence is both a lack of effective strategic planning and an inability to subordinate programs and plans into a coherent system to ensure the implementation of strategic priorities. Such a strategy is yet to be created. It is not just a question of developing another set of documents, but of creating institutional and regulatory mechanisms to ensure that strategies and strategic plans are focused on the implementation of strategic priorities; that strategic plans are materially justified; that strategic goals are supported by setting quantifiable, resource-bearing targets in more specific programs and plans; and finally that institutions are in place to manage the implementation of strategic plans and accountability for achieving strategic goals.

6.2 PRACTICAL EXPERIENCES IN STRATEGIZING REGIONAL AND SECTORAL DEVELOPMENT

6.2.1 STRATEGIZING REGIONAL SOCIOECONOMIC DEVELOPMENT

Subjects of the Russian Federation and municipalities are named in Law No.172-FZ as participants in strategic planning. They (as well as public administration bodies of the federal level) are involved in target-setting, forecasting, planning, and programming for social and economic development.

By Order No. 207-r of the Government of the Russian Federation, dated February 13, 2019 and pursuant to Presidential Decree No. 13 "On the Approval of State Policy Principles of the Russian Federation's Regional Development for the Period until 2025" of 16 January 2017, the following plan was approved: "Strategy for the Spatial Development of the Russian Federation for the Period up to 2025." However, the strategy is methodologically very poor. "In developing any strategy, but especially at the national and regional level, three key tenets must be considered. Any strategy can only propose priorities that reflect national and regional interests and must be fully supported by all kinds of human, material, financial, as well as infrastructural, resources. Only those priorities that offer a competitive advantage are accepted for implementation. But this requires a huge amount of work, identifying where enterprises exist, what labor resources are available in the region or in the city and much more. None of this is in the strategy […] The submitted document is a set of good wishes, and not always useful for the country and its subjects."[7] To illustrate the level of "validity" of this document, here is an example: "The strategy has Appendix 1, which is called the 'List of Promising Economic Specializations for the Subjects of the Russian Federation.' Now, 50 regions have the production of trailers or semi-trailers as one of their most important specializations. It is hard to imagine that two-thirds of the country's regions should be engaged in the production of semi-trailers!"[8] So this document can hardly be called a strategy. And there is a long way to go from adopting a pan-federal strategy to creating effective regional strategies. This requires a mapping of regional development levels. This is emphasized by corresponding member of the Russian Academy of Sciences P. Bakhtizin: "Significant effect in solving this important task can be achieved only within the framework of the implementation of the issues of economic alignment of the regions in the strategy of socioeconomic development of the country, primarily in the system of goals and tools of the policy of spatial regulation."[9]

Regional development strategizing requires solving a rather controversial task. The strategic goals of regional development, on the one hand, should be in line with those at the national (federal) level and localized in a given region. On the other hand, the great diversity of conditions in different regions requires substantially different approaches to regional development strategies. Moreover, according to RAS Corresponding Member Kh. H. Gizatullin, it is necessary "to avoid the temptation of forming a unified model of development, uniform for different levels of economy, forms and methods of organization in the investment sphere."[10] This position is echoed by A. R. Bakhtizin and his colleagues: "We believe that the results of this analysis confirm the validity of the position that *the implementation of a unified regional policy in the Russian Federation is inadvisable.* As we have tried to show, the reasons for differences in the levels of socioeconomic development of the constituent entities of

---

[7] Kvint, V.K; Semi-trailer Territory. Does Russia Need a Spatial Development Strategy?; *Ogonyok* 2019, *No. 10*, P. 8.

[8] *Ibid.*

[9] Bakhtizin, A. R.; Buchwald, E. M.; Economic And Legal Prerequisites And Institutions To Reduce The Level Of Inter-Regional Differentiation In The Socioeconomic Development Of The Russian Federation Subjects; *Journal of Russian Law* 2018, *No. 9*, P. 104

[10] Gizatullin, H. N.; Garipov, F. N.; Garipova, Z. F.; Problems Of Managing Structural Transformation Of The Regional Economy; *Regional Economy* 2018, *V. 14, Issue 1*, 44–45.

the Federation may be found, among other things, *in differences in the development factors of regional economies*. This difference may be the main cause of the gaps in these levels, not the other way around. We also support the position that the mechanisms for equalization of development levels in the federal entities, which have been used for many years, have no future."[11]

The regions act as participants in implementing strategic programs, projects, and plans developed at the federal level. At the same time, as we pointed out earlier, there is essentially no development strategy for the Russian Federation, and the federal-level strategic documents pay too little attention to regional specifics. Therefore, public authorities in the regions must choose goals and objectives similar to those stated in the strategic plans of the federal level as priorities for regional development. However, it is not realistic to use instruments available at the federal level to achieve these objectives at the level of the entities of the Russian Federation, as monetary, customs, and migration policy issues are their exclusive domain. In addition, the fiscal and budgetary policy options at the level of the constituent entities of the Federation (even more so at the municipal level) are considerably limited.

To rely on budget subsidies from the federal center to improve the quality of life on the basis of a high level of socioeconomic development is to perpetuate an abnormal situation in which almost all Russian regions are dependent on such subsidies. Specialists of the Institute of Economic Forecasting in the Russian Academy of Sciences note: "The strategic goal of Russia's spatial development is to transform the country's regions into territories with a high level of economic and infrastructure development, comfortable for people's lives in terms of social and environmental conditions. This goal should be achieved not so much through budget transfers and subsidies but through investments in carefully selected projects that accelerate regions' social and economic development."[12]

At the same time, the practice of reducing the regions' own budget base does not sit well with the targeting of equalizing regional development levels, which in such circumstances becomes an illusion. RAS

---

[11] Valentey, S.; Bakhtizin, A.; Kolchugina, A.; Readiness Of Regional Economies For Modernization; *Federalism* 2018, *No. 3*, 154.

[12] Govtvan, O. D.; Gusev, M.S.; Ivanter, V. V.; Xenophontov, M. Y.; Kuvalin, D. B.; Moiseev, A. K.; Porfiriev, B.N.; Semikashev, V. V.; Uzyakov, M. N.; Shirov, A.; System Of Measures To Restore Economic Growth In Russia; *Problems of Forecasting* 2018, *No. 1*, 3–9.

Academician P. A. Minakir believes that this is a weakness of the law on regional strategic planning, "according to which each subject of the Federation should present its strategy for a 15-year period, and this strategy should envisage a "breakthrough," a "new quality," "new horizons," etc. In this case, no changes are expected in the trends and norms of income distribution between the center and the regions."[13] Thus, the existence of approved strategic planning documents in the subjects of the Federation does not yet turn them into effective tools for implementing strategic objectives. These documents do not constitute a unified system in which the establishment of specific objectives and the identification of the means to achieve them serve an overall goal. Often there is actually no understanding of the connection between strategic planning and long-term plans and the need to create mechanisms for the management and control of the implementation of developed plans and programs and evaluation of their effectiveness.

The analysis clearly shows that the existing regional strategies (as well as the resulting programs, projects, and plans for regional development) were usually not conceived as a system but rather as a list of measures aimed at solving problems perceived as urgent at the time. However, in terms of the means of implementing these projects, the focus has often been on attracting specific investors and the programs have therefore been to some extent "investor driven," while the improvement of the investment climate in the region as a whole has been relegated to the back burner. Due to the low quality of these documents, they are more like declarations of intent, rather than instruments of real strategic management of socioeconomic development.

Among the unfortunate examples is the Siberia Economic Development Strategy, whose objectives have been scientifically well founded and largely retained in the final document, despite objections from federal agencies. The Strategy underwent expert review by the Russian Academy of Sciences and was approved by decree of the Government of the Russian Federation No. 765-r of June 7, 2002. However, its implementation was actually derailed because, although the objectives and resource rationale of the document were quite well elaborated, as noted by RAS Academician

---

[13] Minakir, P. A.; Regional Strategies and Imperial Ambitions; *Spatial Economics* 2015, *No. 4*, 10.

V. V. Kuleshov, "the institutional conditions and mechanisms for implementation of the Strategy were poorly reflected."[14]

The lack of involvement of business and civil society institutions in strategic planning has been convincingly demonstrated in practice, which weakens the social orientation of the program developed and reduces the effectiveness of ongoing monitoring of their implementation.

Under such conditions, the achievement of strategic priorities should be envisaged in the strategic development of the regions.

First of all, it should be emphasized that prior to developing strategic plans, it is necessary to analyze the forecasts of the object's development with a long-term perspective (up to 40–50 years). Foresight methodology can be used in regional strategic planning to assess the potential and priorities for regional development. Such assessments form the basis of the regional development strategic plan.[15] However, not all regions have scientific centers capable of solving the tasks of science and technology forecasting. Corresponding Member of the Russian Academy of Sciences Zh. A. Ermakova points out: "The most difficult task in the process of shaping the concept of scientific and technological development is its technical and technological justification, as institutional structures capable of solving problems of this nature do not exist in all regions."[16] Strategic planning in itself cannot be reduced to the activities of regional authorities and administration. The development and implementation of strategic plans and programs should include representatives of business, nonprofit organizations, and various public associations, up to and including the initiative participation of residents on a case-by-case basis. Maximum openness and the establishment of information channels for interaction between strategic planners allows a variety of interests to be taken into account as far as possible.

An equally important aspect of strategic planning in the regions is to improve the functioning of socioeconomic institutions to create a favorable investment climate, including measures to reduce administrative

---

[14] Kuleshov, V. V.; Seliverstov, V. E.; Strategy for Socioeconomic Development of Siberia: *Institutional Conditions and Implementation Mechanisms* 2005, *No. 4*, 8.

[15] Tretiak, V. P.; Regional Foresight: Possibilities Of Application, Znanie, 2012; *see also*: Foresight "Russia": Design of a New Industrial Policy: *Proceedings of the St. Petersburg Economic Congress (SPEK-2015)*; Bodrunov, S.D. ed. Cultural Revolution, 2015.

[16] *Ermakova J.* Technological Priorities As The Basis For Scientific And Technological Development Of The Region's Industrial Complex // Bulletin of the Orenburg State University. 2012. No. 8 (144). August. P. 107.

barriers to investment activity, eliminate corruption, and ensure reliable legal protection of participants in production activities. In conjunction with the regional strategic planning process, efforts should also be made to harmonize the regional strategic plans with the goals and objectives of the federal strategic planning documents.

Regional strategic plans should be formed considering the indicators of fiscal planning, and compliance of economic and social objectives with the mandatory elaboration of the spatial aspect of the development strategy based on the territorial governance structure of the region. To this end, plans should include the identification of the most promising territorial clusters that can act as drivers of socioeconomic development in the region and work out measures to support them (in economic and administrative aspects).

Since the territorial aspect of development is one of the utmost importance, it is advisable to include a separate territorial development planning document in the package of strategic planning documents for the socioeconomic development of the region. This document should not copy the "Strategy of the Russian Federation for the period until 2025." It should adhere to a common methodological approach to its formulation, taking active account of local conditions and capacities. The Strategic Spatial Development Plan should consider the different scales of territorial clusters at federal and regional levels, the different mandates of federal and regional bodies and the specifics of the regional territorial structure, as well as the involvement of municipalities (taking into account their capacities and coordination of their interaction) in the preparation and implementation of such plans.

The lack of attention to the territorial, spatial aspect of the regional development strategy leads to increasing disparities in territorial development. The rapid formation of the territorial location of production and human settlement, characteristic of the mid-Soviet period of the country's development, has led to the swelling of large urban agglomerations. In the post-Soviet period, urban sprawl developed within cities, and suburbs became densely populated with residential areas without adequate public and social infrastructure. The regional governments have taken little or no notice that most jobs are concentrated in the cities, placing excessive pressure on the transport arteries linking the cities and suburbs. In addition, the concentration of most economic activity in cities makes revenues in peri-urban municipalities relatively scarce, which prevents them from coping with emerging disparities and hampers the resolution of social problems at the municipal level.

Mass migration from small- and medium-sized cities to regional centers creates even more serious problems. The population of many small towns is shrinking, business activity is collapsing, incomes for the rest of the population are falling, and municipal budgets are deprived of the means to maintain the normal functioning of municipal, transport, and social infrastructure, which are rapidly depreciating and falling into disrepair.

Equally problematic is the emerging trend of deepening regional disparities in the development of the agricultural sector. As A.A. Anfinogentova rightly points out, the formation of a strategy for the territorial development of Russia's agro-industrial complex should be aimed at overcoming the increasing differentiation of development levels of agro-industrial sectors under conditions of competition and weakening of the regulatory function of the state, caused by deepening differences in the average per capita real income of the population and the per capita consumption of basic foodstuffs. The existence of regions where social, economic, and environmental problems are particularly acute calls for urgent solutions at the federal level."[17]

The urgent solution to these problems requires a fundamental change in approaches and serious attention to the spatial aspect of strategic planning of regional socioeconomic development. Improving the territorial organization of regional development should aim to overcome imbalances in the formation of territorial economic and settlement clusters in the region. This will make it possible to effectively use the economic potential of different territorial entities, ensuring sustainable socioeconomic development of both individual territories and the region as a whole. Sustainable development planning is connected "with identifying the actual state of the parameters determining the integral result of the system and forecasting their values in the future, i.e., it is mainly connected with the establishment of the system's states. It is crucial to identify the resulting momentum, which is also generated by the interaction of different impulses. If this interaction ensures developing the dynamics along the intended trajectory then the basic characteristics of the system remain unchanged and equilibrium is maintained. This state of equilibrium is stable. Stability means

---

[17] Anfinogentova, A.A.; Use Of The Worldwide Database "Input-Output" to Justify the Development Strategy of Russian Agro-Industrial Complex; *Economics and Management* 2015, *No. 3 (113)*, 9.

the ability of a dynamic system to maintain its motion and function despite perturbations (both external and internal)."[18]

Taking into account the spatial aspect of regional strategic planning will support the development of municipalities in the region's territory. However, this requires a differentiated approach to different territorial formations: small towns, single-industry towns, depressive settlements, and large industrial centers. A careful choice of instruments to influence the development of economic activity in such territories is needed, especially in terms of support for small and medium-sized enterprises.

Territorialization does not mean a one-size-fits-all approach to regional and municipal development (the other extreme). Regional strategic planning should consider the region's place in the national strategy for socioeconomic development, as well as the interdependence and interaction of the federal subjects and the territorial entities within them. This is particularly the case in regions that are closely linked by transport, migration and production links, sharing of natural resources, etc.

Obviously, the regional and federal components of strategic planning are inseparable. Therefore, strategic planning methods are needed to ensure coherence between plans and programs at federal and regional levels and allocate budgets and resources to these plans accordingly.

The effectiveness of regional strategic planning will be achieved when the prepared plans and programs achieve not only the correspondence between the development goals and budgetary/resource endowment. A program of action to *realize* these objectives is also needed (setting up management arrangements for such actions, monitoring how these actions are bringing the objectives closer to realization, and adjusting management decisions).

6.2.2   REGIONAL STRATEGIC PLANNING EXPERIENCES

An example of an attempt at regional strategic planning is the development of the Strategy of Economic and Social Development of St Petersburg for the period up to 2030, approved by Decree of the Government of St Petersburg No 355 of May 13, 2014. One of the authors of this text (V.L. Kvint), as part of the Expert Council for Economic Development under

---

[18] Garipov, F. N.; Gizatullin, H. N.; Stability of Functioning of Production and Economic Systems; *Regional Economy* 2012, *No. 4*, 118.

the Governor of St. Petersburg, and another author (S. D. Bodrunov), who used to head the St Petersburg Committee for Economic Development, Industry and Trade in the rank of a member of the Government of St. Petersburg and was drafting the strategic plan for the development of the city industry until 2020, participated actively in its examination.

The attempt to develop such a strategy has largely reflected the soundness of the approaches taken. At the drafting stage, it was possible to involve not only government representatives, but also qualified experts, academia, universities, and research organizations in the preparation of the strategic development plan for the city. To ensure further active participation in implementing the strategy, not only the city's largest companies and corporations but also companies of federal importance and importantly representatives of various civil society structures were involved in its development.

To better take into account the interests of various segments of the population, the draft strategy was presented for public discussion.

Ensuring the implementation of the St. Petersburg Development Strategy required the creation of appropriate governance mechanisms, as reflected in the Temporary Provisions on the State Planning System of St. Petersburg.

A broad public discussion of the strategy's concept showed that while it was a very progressive step in bringing order to urban planning, it did not fully meet the requirements of this kind of document.[19] For example, a relatively weak point of this document was the lack of a detailed forecast of the city's socioeconomic development, including the use of foresight methodology.

At the same time, some of the ideas underlying the development of this strategy, despite their controversial assessment by the scientific and business communities, deserve attention. We are referencing in particular the concept of a "creative city," similar to the concept of a "smart city."

The attempt to implement these ideas from the perspective of post-industrial societal concepts cannot be considered valid. After all, in this case, the industrial basis of the city is inevitably relegated to a second or

---

[19] Bodrunov, S. D. On the Status of the Strategy for Economic and Social Development of St. Petersburg for the Period Up to 2030; *St. Petersburg Offers* 2017, No. 2(25), 10–11; Karlik, A. E.; Industry As a Structural Element of St. Petersburg Development; *St. Petersburg offers* 2017, No. 2(25), 11–12; Lobin, M. A.; Strategy 2030. Position of the Union of Industrialists and Entrepreneurs of St. Petersburg; *St. Petersburg Offers* 2017, No. 2(25), 20–21.

third role, and there are essentially no serious prospects for it. As such, the strategy does not suggest active and, more importantly, purposeful participation of the city in the transition to the sixth technological mode, in the burgeoning next technological revolution, and in achieving country's potential within the perspective of transition to NIS.2.

Nevertheless, the components of these concepts themselves can find useful application if they are not based on the contrived post-industrial status of the region but the real potential contained in the technological and economic revival of industry, i.e., in the reindustrialization of the city economy based on an advanced technological way of life.

The creative side should not be achieved by destroying industrial capacity and creating a "creative environment" out of its ruins. It is the technological modernization of industry that should become the "creative environment" that ensures progress in the socioeconomic status of the city, including the development of science and education. Then the creative elements will not be developed for the sake of creativity "in general," but for concrete results that strengthen the high-tech industry sectors that serve the economic growth of the city and the well-being of its inhabitants. In this context, the focus on human development as the main driver of socioeconomic progress is a realistic one.

This approach is supported by Academician of the Russian Academy of Sciences V. V. Okrepilov, who states: "The development of a "knowledge economy," including education, science, health, information and biotechnology, innovative industries, and the creation of new knowledge and technologies in all fields of activity is becoming a major focus. According to the Strategy, the "knowledge economy" should be St. Petersburg's top priority, its greatest contribution both to the development of Russia as a whole and to the well-being of St. Petersburg residents, a significant proportion of whom will eventually be employed in high-tech labor."[20]

This in no way contradicts the status of St. Petersburg as the cultural capital of Russia. Furthermore, cultural progress is a prerequisite for people-centered economic development and makes an invaluable contribution to achieving the primary goal of social production: human development. There will therefore be a growing worldwide interest in cultural heritage, which, even from a utilitarian-economic point of view, is an

---

[20] Okrepilov, V. V.; Quality of Life: Guidelines for Strategy 2030; *Quality Economy (Online)* 2017, *No. 1 (17)*. www.eq-journal.ru http://eq-journal.ru/pdf/17/%D0%9E%D0%BA% D1%80%D0%B 5%D0%BF%D0%B8%D0%BB%D0%BE%D0%B2.pdf

effective investment. In this context, the concept of noonomy assumes simultaneous, synchronized development of the industrial-technological progress (1) and the cultural factor (2) for a genuine transition to NIS.2, achieved exclusively as a space of socialized solutions in the economy, facilitated by the first factor, and fostering the populations' nonsimulative consumer behavior, which is the task of the second. At the same time, such elements of the "creative city" concept as the creation of a city with a high quality of life oriented toward real needs and the requirements of people can be implemented only if the city has a solid foundation not only in the sphere of services but also in the sector of modern material production.

"The Smart City is intended to create a citizen-friendly urban management system that responds promptly to the needs of the population and ensures accessibility and high quality of public and social services. Its other components, including transforming the living environment and improving the quality of life (high accessibility to education, health care, and social services), can also only be built on solid industrial foundations, not on post-industrial mirages.

It is no coincidence that the model of the modern smart city includes as a fundamental element the smart economy based on high-tech industries organized on the principles of smart manufacturing, which is supposed to be the basis of Industry 4.0, created during the unfolding industrial revolution.[21]

It creates a high-tech industry that gives the city well-paying jobs that require quality education and high qualifications that is a breeding ground for innovation, which will become the field of activity for a genuine "creative class." "Creativity" aimed at creating surrogates for art or media gum, even if it sells well, cannot be the basis for the development of a smart city.

The presence of a layer of highly educated and skilled citizens is the basis for the active participation of residents in the strategic planning of the city. They are able to demonstrate both civic engagement, interest and the ability to come up with constructive proposals. For them, the practical

---

[21] Giffinger, R.; Gudrun, H.; Smart Cities Ranking: An Effective Instrument For The Positioning Of Cities?; *ACE: Architecture, City & Environ* 2010, *vol. 4, issue 12*, 7–25; Meijer, A.; Bolívar, M.; Governing The Smart City: A Review Of The Literature On Smart Urban Governance; *International Review of Administrative Sciences, Volume 82, Issue 2*, 392–408; Caragliu, A.; Del Bo, C.; (2012) Smartness and European urban performance: Assessing The Local Impacts Of Smart Urban Attributes; *Innovation: The European Journal of Social Science Research* 2012, *vol. 25, issue 2*, 97–113.

exercise of the right to participate in decisions affecting their city is a social value in its own right.

One of the most eminent figures in contemporary geography, David Harvey, stressed that nowadays citizens' right to their city could not be reduced to individual or group access to urban resources. It becomes a right to change and renew the city according to one's heart, a right that is more collective than individual.[22]

The most important factors for good strategic planning are monitoring not only the implementation of the strategy, but also the changes taking place in the object of strategy, its internal and external environment, and adjusting the strategy accordingly.

The strategic concepts of regional development (especially the plans and programs that specify them) should have clear criteria for evaluating their implementation in terms of the effectiveness of the solutions applied and their impact. For example, Academician B.N. Porfiryev proposed such a system of evaluation criteria for the state program for the development of the Far East region.[23]

A city, especially as a constituent entity of the Russian Federation (like St. Petersburg) and in that sense as a region, is in a state of continuous development. The urban production environment and infrastructure, internal and external economic conditions, social-demographic structure, interests, and preferences of its inhabitants are changing. All this requires regular adjustments to the strategic plans of regional development. Therefore, the adopted strategic programs cannot remain unchanged. The framework for the concept of strategy tagging contains the provision that if the main strategic goals are unchanged (otherwise it is not a strategy) many aspects of the strategic plan can and should be adjusted to increase the viability of the strategy and the effectiveness of its implementation.[24]

The elaboration of strategic plans for the development of the region faces contradictions related to the limited economic factors of development. At the same time, it would be unrealistic to try to implement an overly broad range of priorities. If at the given moment there are not

---

[22] Harvey, D. *Rebel Cities: From The Right To The City To The Urban Revolution*, Verso, 2012, 4.

[23] Lexin, V. N.; Porfiriev, B.N.; Assessment Of The Effectiveness Of State Programmes Of Socioeconomic Development Of The Regions Of Russia; *Problems of Forecasting* 2016, No. 4 (157), 86–87.

[24] Kvint, V.; *A System Of Principles For Strategic Planning: The Concept Of Strategizing*, RAS-HSIU, 2019, 104–105.

enough resources for the implementation of any priority, or if this priority does not ensure the growth of competitive advantages of the region, then this priority should be excluded from the strategic plan. This sometimes makes it impossible to implement even potentially promising and effective projects. However, strategizing is the tool to overcome this contradiction by focusing on capacity building for the basic factors of economic development.

An example of such an approach to regional strategic planning is provided by the Strategy for the Development of the Maritime Region. To ensure a balance between the interests of different social groups and the resources required to satisfy them, the strategy includes a block that ensures the removal of economic constraints to development. The drafters of the strategy (one of the authors of the book was among them) designed it based on a combination of three blocks (three models according to the strategy's terminology), each of which addresses a different group of problems.[25]

The first block aims to remove infrastructural and institutional constraints to economic growth. The second is responsible for the active support of the regional authorities and administration for the development of priority industries and production complexes.

Finally, the third block should provide the innovation component the creation of new products and new industries to form new points of growth.

This construction of a regional development strategy for Primorsky Krai creates certain difficulties in project management. The objectives of the second and third clusters can only be effectively achieved if, at least in part, the objectives of the first cluster are achieved. However, the high volatility of global markets, which also significantly impacts the domestic market, makes it necessary to rush to adopt new technological solutions, new product lines, and new business models.

The development of a regional strategy should be based on the identification of the region's specificities and the search for its competitive advantages, which enables the identification of promising points of growth.[26] In this sense, the peculiarity of Primorsky Krai is its proximity to

---

[25] Darkin, S. M.; Strategic Problems of the Russian Far East; *Management Consulting* 2016, *No. 1*, 70.

[26] Novikova, I. *The Russian Far East: Strategic Development of the Workforce*, Apple Academic Press, 2020.

China, which represents a huge market, but at the same time, a competitor with a progressively developing economy.

China's development is based on strategic planning, with very far-reaching goals. The horizon of the PRC strategy covers the period 2012–2050. In addition, indicative parameters of the strategic plan are designed for different planning stages in annual, quinquennial, and decennial increments.[27]

To ensure a high level of competitiveness in such conditions, it is necessary to focus on the most advanced production technologies. And by no means should we focus only on the import of technology. Without attention to financing domestic technological developments, not only in recognized domestic science and technology centers but also in regional research and development clusters, the conditions for sustainable economic growth cannot be created. It should be based, among other things, on its own scientific and technological core of the regional economy.

For example, the Far Eastern Federal District has good prerequisites for efficient mariculture business. Mariculture technologies are widely used in the neighboring countries of Japan, China, and Korea, whose experience could be used in Primorsky Krai. In addition, the National Research Center for Marine Biology in the Far Eastern Branch of the Russian Academy of Sciences operates in Vladivostok, a major research center whose research facilities can be used to develop rational methods of mariculture management in the region.

Balanced regional development in modern conditions requires special attention to large urban agglomerations. Many regional centers have become such agglomerations, and Moscow, St. Petersburg, and Sevastopol are separate entities of the Russian Federation.

Agglomeration growth in our country, as elsewhere in the world, is caused by the tendency of businesses and the population to gravitate toward centers with a developed transport, public, and business infrastructure, and with a high concentration of scientific, educational, and cultural institutions. However, as we noted above, the spontaneous growth of urban agglomerations creates internal imbalances and tensions and slows down smaller territorial formations' development. So far, there has been no real planning and control of urban agglomeration growth, and strategic plans for regional development must fill this gap.

---

[27] Muratshina, K. G.; China-2050: Specifics Of Strategy Formation; *Izvestia of the Ural State University* 2010, *No. 3 (80)*.

During the development of strategies for regional socioeconomic development, a system of principles for strategic planning of regional development was formulated (according to V.L. Kvint's methodology of strategic planning):

- prioritizing the national interests of the Russian Federation as determined by strategic planning documents;
- recognizing the special role of human beings and knowledge in socioeconomic development in the context of contemporary challenges;
- ensuring the sustainability of development based on the balance of socioeconomic interests;
- mandatory application of indicators for assessing the level of achievement (performance) of the established objectives;
- aiming to achieve the best possible results and taking all necessary measures to do so;
- considering resource opportunities when choosing the main priorities and goals of socioeconomic development;
- ensuring effective functioning of state and civil society institutions."[28]

6.2.3 INDUSTRY STRATEGY

Sectoral strategy is defined by a general vision of the strategic outlook and development goals defined by achieving the concept of economic reindustrialization and a vision of the contribution of specific sectors to this task.

As in the case of regional development strategy, a complex network of sectoral interdependencies needs to be considered. In addition, sectoral strategic planning cannot ignore the spatial aspect of the industry, which determines the importance of coordinating sectoral and regional strategic planning.

The vision of the national socioeconomic development strategy developed here, which includes the concept of the Russian economy's reindustrialization based on high technology within the framework of the

---

[28] Bogdanova, N.V.; Fieraru, V.; Features Of Strategic Planning And Development Of Competitive Advantages Of Urban Agglomerations (On The Example Of St. Petersburg); *Management Consulting* 2017, *No. 2*, 123.

transformation of the global economy to a new type, defines the target setting for modernization of both individual sectors and the sectoral structure of the Russian economy as a whole. In doing so, the mobilization of intra-industry development capacities should contribute to the resolution of priority tasks aimed at the technological modernization of the economy at the most advanced level.

It should be stressed that priority should be given to strategic decisions aimed at implementing domestically developed technologies.[29] Only this approach ensures real growth of the country's own technological culture and the formation of a national scientific-technological core. Unfortunately, the opposite approach is quite common. A striking example: the national project to create a high-speed electric train Sokol (ES-250) commissioned by RAO VSM was once rejected in favor of buying the Sapsan train from Siemens.[30] The reasons seemed to be quite good a number of components of the developed train did not provide a sufficient level of reliability, and the development itself was delayed.[31] However, the leadership of the Ministry of Railways was faced with a choice either to bring the domestic development to the required technological level and thus create a new growth point for high-tech production in the country or to finance the development of high-tech production abroad. This choice is not a sign of strategic wisdom; instead, the national development strategy has been sacrificed for tactical corporate gains (with the company owner—the state—being inattentive to this issue).

Corresponding Member of the Russian Academy of Sciences I.I. Eliseeva I. states, "Acquiring the results of others' R&D (licences) means acquiring some important dynamic resources, but does not mean developing your own organizational capabilities. It is also worth recalling another principle of the resource-based approach: to build a sustainable competitive advantage, a resource must be unique. It is difficult to count on the uniqueness of the knowledge acquired."[32]

---

[29] Kvint, V. *A System Of Principles For Strategic Planning On the Choice of Priorities*; Budget 2016, No. 11, 78–81.

[30] Razumeeva, V.; Railways Are Gaining Speed; Business Guide (Railway Transport), *Kommersant* 2009, *No. 239 (4294)*. http://www.kommersant.ru/doc/ 1292160

[31] Guryev, A. I.; And Why Russians Didn't Like to Drive Fast? The Story of a Doomed Project. OOO Izdat.-poligraf KOSTA, 2009, 206–217.

[32] Eliseeva, I. I.; Platonov, V. V.; Dynamic Potential an Understudy of Innovation Activity; *Finance and Business* 2014, *No. 4*, 107.

It should be borne in mind that technological progress in each sector produces a synergistic effect, stimulating technological innovation in related sectors and thus ensuring the development of the agricultural sector as a whole. On the contrary, the slowdown of technological progress, technological backwardness, and even more so, technological degradation, pulls back the entire national economy. Industry strategizing has so far often ignored these obvious patterns.

Another shortcoming is that when setting objectives for sectoral (as well as regional) development strategies, different government departments do not integrate such strategic plans and programs in a meaningful way, starting with the methodology of their development. This situation is an inevitable consequence of the lack of a well-thought-out development strategy for the Russian Federation.

Another shortcoming of sectoral strategies is the gap between the resource endowment of projects assumed in strategic plans and programs and the actual mechanisms for allocating the corresponding funds through budgetary and credit planning. Calculations by the sectoral agencies are not implemented through the respective decisions of the Ministry of Finance and the Central Bank, either because of a mismatch between the requests and the possibilities or because of the different visions of the necessity of allocating certain funds from the point of view of the different state bodies.

Often in documents called sectoral strategies, there is no clear understanding of either the strategic goals or how to achieve them.

The example of Russia's food market strategy shows how approaches to sector strategy are evolving and how the problems of sectoral development stratagem are being addressed.

Food production in our country is a huge industry that needs to be addressed in several ways. One of the most pressing issues is ensuring food security in the Russian Federation. The escalation of Russia's relations with the US and the EU has had several political and economic consequences (various sanctions, etc.) that have negatively impacted the Russian food market. The need for an import substitution policy has emerged, and a strategic approach to developing domestic food and agricultural raw materials production has become a priority. Without this, it is unfeasible to achieve the overall strategic goals of Russia's development through the priority improvement of the "quality of life of Russian citizens by guaranteeing high standards of life support,"[33] including not only the provision of a balanced diet but also progressive changes in the lifestyle of rural toilers.

A sectoral strategy for food production can only be implemented as part of a national strategy that is also coordinated with the development strategies of other sectors that supply the food sector with resources (machinery, equipment, fertilizers, electricity, etc.) and that promote food products to the end-user (transport, storage, processing, wholesale, and retail). These inter-sectoral linkages should be considered in the food development strategy along with regional ones. The territorial peculiarities of food production, which depend on regional natural and climatic features, play a significant role in developing this sector.

When formulating the sectoral development strategy, it is critical not only to clearly justify and formulate the goals and objectives that are translated into strategic programs and plans, but also to give them a verifiable expression in the form of quantitative and qualitative criteria that are formulated in a certain way. This is necessary for continuous monitoring of the strategy's implementation by the public administration bodies responsible for the program. Thus, the import substitution objective of Russia's food security strategy is important not only in terms of supplying the country with food, but also in terms of keeping the agrarian sector functioning and ensuring sufficient investment for its technological modernization.

In implementing these tasks, certain successes were achieved in Russia: increasing the share of domestic products on the domestic food market and stimulating the development of the domestic production of agricultural machinery and equipment. There has been a partial recovery of agriculture after the recession of the 1990s.

The regional aspect of the food industry's strategic development should be focused not only on account of local climatic conditions and production specialization in the region but also on the study of the local food market structure in terms of analysis of opportunities to meet regional demand.

The legal protection of food market participants (especially peasants and farmers) and the economic infrastructure for marketing products and inputs to those households remain a challenge for the food market strategy.

---

[33] Russian Federation Food Security Doctrine; Portal of the President of Russia, February 1, 2010. http://kremlin.ru/events/president/news/6752.

Tourism is a promising sector in terms of its contribution to the overall socioeconomic development of the Russian Federation. This prospect was apparent in connection with the Covid crisis of spring–summer 2020. In Russia, as the analysis has shown, there is considerable potential for tourism development in various regions of the country. Exploiting this potential would make it possible to boost the efficiency of the economies of these regions and provide competitive services based on low-demand cultural heritage sites. The original archaic phenomena of national cultures are also attractive in terms of the tourism business.

Strategic planning of the tourism and recreation sector is designed to mobilize this development potential, which was underutilized in the previous decades. The structure of domestic demand for tourism and recreation services was characterized not only by a general decline in demand in the 1990s but also by a bias toward imported services.[34] Low demand, especially for domestic tourism, led to a decline in the hospitality industry and a high level of deterioration of communal and transport infrastructure in the tourist areas.

A significant part of tourist service facilities and the hospitality industry has degraded, leading to a sharp differentiation in their comfort level and facilities. This differentiation corresponded to the social polarization of the population, many of which could not afford high-quality tourism services and comfortable accommodation.

For a long time, the tourism industry was not considered as an object of state regulation, let alone strategic planning. It was assumed that creating a competitive market environment was not enough for a balanced development of the industry. In fact, nothing has been done to restore mass domestic demand for tourism services. In terms of attracting foreign tourists, reliance has been placed exclusively on private initiative of the tourism business.

Massive asset retirement in the tourism and recreation industry has been interpreted as a natural process of adjustment to market conditions.[35]

However, the past crisis has shown that this sector should be included in the strategic directions of the state economic development. These objectives have been set in two-state programs to develop the tourism sector (up

---

[34] Desyatnichenko, D.; Desyatnichenko, O. Yu.; Theoretical Aspects Of Forming the Strategy of Recreation and Tourism Development in the Region; *Management Consulting* 2016. *No. 4*, 153.

[35] *Ibid.*, 154.

to 2020 and up to 2035).[36] Under current conditions, strategic development of the tourism sector should be based on identifying the factors that form the attractiveness of tourist services and assessing the tourism and recreational potential of various sites and territories. There is a need for long-term forecasting of the role of the tourism industry in the overall sustainable development of the economy.

Strategic planning is designed not only to set goals for the industry but also to find mechanisms to achieve them, primarily by activating innovative processes that ensure the application of advanced organizational and technological solutions to make the market of tourist services possible more attractive to domestic and foreign consumers.

From this point of view, it is necessary to focus on the whole set of factors that can improve the quality of services for tourists. There is a growing importance of investment in transport infrastructure, which increases the accessibility of tourist sites; in the improvement of facilities and the level of organization of hotel services. Tourism development also faces environmental constraints and not only in the zone of specially protected areas. These problems are everywhere from rural provinces, where local ecosystems can become unbalanced, to large urban agglomerations, where traffic increases pollution.

The spatial and territorial strategy for the development of the tourism industry involves the regulation of the location of tourism infrastructure facilities, taking into account the presence of zones of permanent residence of the population and zones of industrial and production activity. The development of these areas should be planned based on the presence of cultural and natural objects in these areas, which are points of attraction for tourist flows.

Therefore, in tourism and recreation policy-making, a cluster approach should be used, which allows the scale of the anthropogenic pressure on the area and the associated risks to be balanced against the benefits for the economic actors involved in the tourism industry.

It is vital to combine tourism services of different specializations in a rational and complementary way to create synergies—the greater the diversity of supply on the tourism market, the higher the demand will be. The complexity of tourism services requires an appropriate approach to their standardization. J.A. Ermakova draws attention to this:

---

[36] For an overview and comparison of the targets of these programs, *see* Bushueva, I. V.; New Strategic Priorities Of Tourism Development In Russia; *Service Plus* 2019, *V. 13, No. 4*, 27–29.

"One of the tools to ensure the quality of tourism products and services is the development and implementation of a set of standards of hospitality in a tourist destination. The complexity of the tourism product implies the inclusion of a variety of services: transport services, accommodation and catering services, entertainment services, etc. All of these areas need to be reflected in one way or another in the set of standardization and certification processes."[37]

In assessing the synergy between the tourism and hospitality industries, we note that their interconnectedness and complementarity create good prospects for integrated spa treatments as an independent sub-sector. The global market for these services is growing rapidly, and Russia still occupies a respectable place, despite the considerable destruction of the Soviet system of spa treatments. By organizing effective preventive and rehabilitative services, sanatorium and spa services can become one of the ways to make the industry more competitive and develop at a high rate in terms of the global and national tourist services market.

The implementation of such priorities for the development of the tourism and recreation sector requires a well thought-out application of strategic principles with strategic plans and programs linking sectoral and territorial planning, and coordinating the achievement of federal, sectoral, regional and local objectives. This should ensure preservation and modernization, plus (potentially) the development of tourism facilities (both natural and cultural) and the improvement of the whole system of tourism infrastructure, including its human resource base. All this is intended to strengthen the industry's position in global, national, and regional markets.

However, all efforts to develop the industry's material and human resources will not be sufficiently effective without proper attention to the market promotion of tourism and recreational services. Unfortunately, little attention has been paid to the marketing side of the industry for a long time, which has held back demand for tourism services in Russia, both nationally and globally. The industry's potential is still not fully exploited, limiting its contribution to the country's overall economic development.

Increasing the quality of tourism services requires a competitive market, which requires the participation of organizations of various forms of ownership and legal form. Mechanisms to stimulate small and

---

[37] Polyakova I. L., Ermakova J. To Standardisation In The Regional Hospitality Industry: Directions, Main Stages; *Bulletin of Orenburg State University* 2015, *No. 8 (183),* 116.

medium-sized businesses, whose significant share creates a competitive environment, should be applied more widely in the industry.

Strategizing this sector involves increasing private investment in the development of relevant tourism and recreation clusters (TR-clusters). However, they are hampered by the lack of a well-developed legal and regulatory framework for the interaction of market players, state regulation's boundaries and rules of tourism activities, and the protection of both tourists and tourism organizations from the risks that may arise.

The strategy of forward-looking services sectors, which include not only tourism but also multidisciplinary services, is designed to enhance the material, cultural, and emotional quality of people's lives, which contributes to the development of creative activity.[38] This is a very significant task as part of increasing the socialization of the economy. However, economic development should be based on the basic sectors of material production; strategizing such sectors is a fundamental task of public administration.

For example, serious problems are related to the strategic development of one of the most important industries of the national economy: the machine tool industry. The industry experienced a deep decline throughout the 1990s and during a decade and a half of this century. Only in the last 5 years has there been a slight recovery, which can only be described as ephemeral given the needs of the national economy for domestic production of machine tool products.

In 2014–2016, the share of imported metalworking equipment in supplies to the Russian market exceeded 90%. After 2016, production of domestic machines began to grow and now, according to the Stankoinstrument Association, the share of imports is around 75–80%.[39] These estimates are very approximate since there are no statistics on the import of metalworking machines.

Another reason is the neglect of the industry, underestimation of its strategic importance for the development of the national economy, and the poor quality of the drafting and implementation of strategic programs for the industry, which are long overdue.

The development of the machine tool industry is a major factor in the successful technological modernization of the Russian machine-building

---

[38] Phelps, E. S. *A Good Economy for China*. https://www.projectsyndicate.org/commentary/china-innovation-good-economy-by-edmund-s-phelps-2016-06.

[39] Tolstoukhova, N.; The Machine Requires Benefits; *Rossiyskaya Gazeta*, October 28, 2018. https://rg.ru/2018/10/28/minpromtorg-sprognozirovalrost-obema-proizvodstva-v-stankostroenii.html

industry, which determines the technological level of the entire national economy.

The industry development strategy should be aimed at recreating the complex of production facilities, providing the formation of the modern machine-tool industry. Everything must be rebuilt, from the casting of blanks and production of components (sub-chips and program control systems) to machine assembly.[40] Such a strategy must be integrated into the overall development strategy of the machine tool industry, the main customer for machine tools.

Unfortunately, the machine tool industry strategy adopted in 2017 does not set goals in this way and, therefore, cannot be implemented. The machine tool subprogram (2011) also did not include such objectives and was not resourced either. However, it contained some objectives for the development and production of machine tool equipment; a time frame for the achievement of these objectives was defined, and the allocation of funds was linked to these specific objectives. One of the disadvantages of the subprogram was the lack of connection between the development of new machine tool products and the needs of potential customers, which jeopardized the series production of these machines. The development was financed by the state and not by individual companies, making the new designs unnecessary.[41]

Unfortunately, the new industry development strategy to 2030 also deviates from scientific principles of strategic planning. The priorities are not explicitly stated. According to the program, this is to ensure Russian companies' leadership in the domestic market and technological security.[42]

The first priority is too ambitious for the resources allocated to it and is not supported by concrete goals and a program of action to achieve them. Taking into account the identified priorities, the strategy identifies three strategic objectives:

1. Increasing the share of Russian products on the domestic market to 50% by 2030;

---

[40] Tkachenko, S. S.; On The Development Strategy Of The Domestic Machine-Tool Industry Until 2030 From The Perspective Of Blank Production; *Metalurgy of Mechanical Engineering* 2019, *No. 5*, 2–4.

[41] Mechanic, A. We Can't Do Without Our Worms; *Expert* 2014, *No. 37(914)*. https://expert.ru/expert/2014/37/bez-svoihchervyakov-ne-obojdemsya/.

[42] Strategy For the Development of the Machine Tool Industry Until 2030. Portal of the Ministry of Industry and Trade of the Russian Federation, 2017, 55.

2. Ensuring the growth of Russian production at an average rate of at least 15% per year;
3. Organizing the competitive production of key components and tools."[43]

One would expect these objectives to be deployed when setting specific targets. But the indicators used in target-setting essentially double down on the formulation of targets, defining only production volumes and domestic market share.[44]

The strategy does not define technological priorities and areas of import substitution; there are no targeted programs aimed at developing the production of high-tech components for the industry, and no program of activities aimed at the development of the machine tool industry. All the specifics come down to references to developing an industry roadmap.

Another problem that the 2017 strategy states, but does not propose a solution to, is the unfavorable financial environment for the development of machine tool production. The industry is not profitable, and with current interest rates, neither short-term loans for working capital nor long-term loans for production development are available to machine tool builders. There is a high level of depreciation of fixed assets in the machine tool industry, which makes it difficult for products to meet global competitiveness standards. The solution to the problem lies in long-term financing.

Another major challenge is the revival of the national machine tool design school, without which technological independence is impossible, either in the machine tool industry or in mechanical engineering.[45]

Nevertheless, even under such difficult conditions, Russian machine tool builders manage to produce high-tech products, some of which are exported. In doing so, domestic producers have to give way to foreign competitors, often not because of the technical level of their products but because of the financial conditions for marketing them.[46] Russian companies (unlike their foreign competitors) cannot provide long-term

---

[43] *Ibid.*

[44] *Ibid.*, 73–76.

[45] Mechanic, A.; Machines For Children And Grandchildren; *Stimulus* 2017. https://stimul.online/articles/sreda/stanki-dlya-detey-i-vnukov/

[46] Zubkova, E.; *Russian Machine Tool Industry: A Thorny Path To Success*, All Industrial Regions of Russia, 2017.

hire purchases or sell products on leasing terms, as this requires long-term unaffordable (for them) bank lending.

Staffing of the industry is also an extremely serious problem. The training of specialists in machine tool engineering has significantly decreased together with the contraction of production in the industry itself. Strategic decisions on the development of the machine tool industry would have to include the development of the training system as a priority. However, while mentioning the human resources issue among the risks to the development of the sector, the strategy is limited to the need to address it.

The implementation of the "Strategy for the Development of the Machine Tool Industry to 2030" is ahead of schedule. This, however, does not remove all the problems associated with the development of the sector, for without a strategic vision and a long-term program of action, it is impossible to achieve the end results stated in the strategy. Therefore, using scientifically based methods and forms of strategic development of the machine tool industry remains an important task.

Conclusion

Any strategy aims to improve the quality of human life in all its aspects. Quality of life is a multifaceted category, and in this book, we have combined two scientific fields in its study: strategy as a science and noonomy as a new scientific perspective on the processes of interconnection between humans and society, and between humans and nature. The chosen approach allows the reader to take an integrated look at how strategy and strategy methodology contribute to the realization of the main categories of noonomy in the qualitative transformation of society and its productive forces.

This book is essentially an outline and sketch of how the processes of strategy change all aspects of human life—its productive, cultural, and spiritual experiences. In many ways, the strategist's goal-oriented activity, which is not prone to economic determinism, contributes to the new character of the interaction between humans and nature predicted by the great scientist Vladimir Ivanovich Vernadsky. His understanding of the human impact on nature is largely comparable to the scale of geological processes. On the other hand, the term "noonomy," complementing the noosphere view of Academician V.I. Vernadsky, reinforces all aspects of people's cultural and spiritual life in strategic processes. The authors have not only tried to present the conceptual positions of the theory of strategy and noonomy, but also to propose to economists, sociologists, cultural studies experts, and even philosophers, a new perspective on the interconnected study of humans, their creative activity, society, and the natural environment.

Bibliography

Aganbegyan, A. G. *Human Capital and its Main Component the Sphere of "Knowledge Economy" as the Main Source of Socioeconomic Growth*; No. 4; Economic Strategies, 2017.

Amin, S. *Russia: A Long Way from Capitalism to Socialism*, Scientific Editor S.D. Bodrunov; INID, Cultural Revolution, 2017.

Amin, S. *October Revolution 1917: A Century Later*, Scientific Editor S.D. Bodrunov; INID, Cultural Revolution, 2018.

Anfinogentova, A. A. The Use of the Global Database "Input-Output" to Justify the Development Strategy of Russian Agro-Industrial Complex, *Econ. Manage.* **2015**, *3*(113), 4–10.

Anfinogentova, A. A.; Yakovenko, N. A. Theoretical and Methodological Problems of Innovative Development of Agro-Food Complex of Russia. *Reg. Econ.* **2011**, *4*, 87–109.

Arendt, H. *Vita Activa, or on the Active Life*; Aletheia, 2000.

Balandin, R. K. *Geological Activities of Mankind: Technogenesis;* Higher School, 1978.

Bakhtizin, A. R.; Buchwald, E. M. Economic and Legal Prerequisites and Institutions to Reduce the Level of Inter-Regional Development of the Subjects of the Russian Federation. *J. Russ. Law* **2018**, *9*, 102–112.

Baudrillard, J. *Pour Une Critique De L'économie Politique Du Signe*. Editions Gallimard, 1972.

Berle, Adolf A.; Gardiner, C. *Means: The Modern Corporation and Private Property*; The Macmillan Company, 1932.

Bogdanova, N. V.; Fieraru, V. Features of Strategic Planning and Development of Competitive Advantages of Urban Agglomerations (On the Example of Saint Petersburg). *Manag. Consult.* **2017**, *2*, 121–127.

Baudrillard, J. *Towards A Critique of the Political Economy of the Sign*. Academic Project, 2007.

Bodrunov, S. D. *Analysis of the State of Domestic Machine-Building and the Imperatives of New Industrial Development*; Institute for New Industrial Development (INID), 2012.

Bodrunov, S. D. Coming and Thinking. *Econ. Reviv. Russ.* **2016**, *4* (50), 11–19.

Bodrunov, S. D. Coming: The New Industrial Society: Resetting, 2nd Ed., Revised and Enlarged; S. Yu. Witte, 2016.

Bodrunov, S. D. Innovative Development of Industry as a Basis for Technological Leadership and National Security of Russia. *Proc. Free Econ. Soc. Russ.* **2015**, *3*(192) 24–56.

Bodrunov, S. D.; Integration of Production, Science and Education as the Basis for ReIndustrialization of Russia. *World Econ. Int. Rel.* **2015**, *10*, 94–104.

Bodrunov, S. D. On the Issue of Noonomy. *Econ. Reviv. Russ.* **2019**, *1*(59), 4–8.

Bodrunov, S. D. On the Issue of Reindustrialization of the Russian Economy. *Econ. Reviv. Russ.* **2013**, *4* (38), 4–25.

Bodrunov, S. D. On the Issue of Reindustrialization of the Russian Economy In the WTO Environment. *Econ. Reviv. Russ.* **2012**, *3* (33), 47–52.

Bodrunov, S. D. What Kind of Industrialization Does Russia Need? *Econ. Reviv. Russ.* **2015**, *2* (44), 6–17.

Bodrunov, S. D. Technology Convergence A New Basis for Integration of Production, Science and Education. *Econ. Sci. Mod. Russ.* **2018**, *1*, 8–19.

Bodrunov, S. D. Modernization of Defense-Industrial Complex and Provision of State Security. *Year Planet* **2005**, *14*, 107–112.

Bodrunov, S. D.; New Industrial Society of the Second Generation: Rethinking Galbraith. *Galbraith: The Return*, Bodrunov, S. D., ed.; *Cult. Revolut.* **2017**, 27–71.

Bodrunov, S. D. New Industrial Society. Production. Economics. Institutes. *Econ. Reviv. Russ.* **2016**, *2*(48), 5–14.

Bodrunov, S. D. *New Industrial Development of Russia in the WTO Environment: Examination of the Adopted Concepts of Innovative Development of Russia*; Institute for New Industrial Development (INID), 2012.

Bodrunov, S. D. *Noonomy: Monograph*; Cultural Revolution, 2018.

Bodrunov, S. D. Noonomy: Ontological Theses. *Econ. Reviv. Russ.* **2019**, *4*(62), 6–18.

Bodrunov, S. D. *Noonomy: Trajectory of Global Transformation: Monograph*; INID, Cultural Revolution, 2020.

Bodrunov, S. D. On Some Issues of Evolution of the Economic and Social Structure of the Industrial Society of the New Generation. *Econ. Reviv. Russ.* **2016**, *3*(49), 5–18.

Bodrunov, S. D. On the Status of the Strategy for Economic and Social Development of St. Petersburg for the Period up to 2030. *St. Petersburg Offers* **2017**, *2*(25), 10–11.

Bodrunov, S. D. *General Theory of Noonomy*; Cultural Revolution, 2019.

Bodrunov, S. D. From ZOO to NOO: Man, Society and Production in the Conditions of New Technological Revolution. *Prob. Philos.* **2018**, *7*, 109–118.

Bodrunov, S. D. The Transition to the New Industrial Society of the Second Generation: The General Cultural Dimension. *Econ. Reviv. Russ.* **2017**, *1 (51)*, 4–11.

Bodrunov, S. D. Reindustrialisation. "Round Table" In the Free Economic Society of Russia. *The World New Econ.* **2014**, *1*, 11–26.

Bodrunov, S. D. Russian Economic System: The Future of High-Tech Material Production. *Econ. Reviv. Russ.* **2014**, *2*(40), 5–16.

Bodrunov, S. D. *Theory and Practice of Import Substitution: Lessons and Problems;* S.Yu. Witte, 2015.

Bodrunov, S. D. *Formation of Russia's Reindustrialisation Strategy*; Institute for New Industrial Development (INID), 2013.

Bodrunov, S. D. *Formation of Russia's Reindustrialisation Strategy*, 2nd Ed., Revised and Supplemented. In Two Parts. Part One; INID, 2015; Part Two, INID, 2015.

Bodrunov, S. D.; Grinberg, R. S.; What to Do? Imperatives, Opportunities and Problems of Reindustrialisation. *Mat. Scientific and Expert Council Under the Chairman of the Federation Council "Reindustrialisation: Opportunities and Limitations," of the Federation Council of the Russian Federation*, 2013, 14–25.

Bodrunov, S. D.; Galbraith, J. K. *Concept of the New Industrial Society: History and Development*, Bodrunov, S. D., ed.; USEU, 2018.

Bodrunov, S. D., Galbraith, J. K. *New Industrial Revolution and the Problems of Inequality*, Bodrunov, S. D., ed.; G.V. Plekhanov Russian University of Economics, 2017.

Bodrunov, S. D.; Rogova, E. M.; On the Basic Principles of Formation of Import-Substituting Industrial Policy in Russia. *Actual Probl. Econ. Manage.* **2014**, *4*(4), 7–12.

Bodrunov, S. D. Lopatin, V. N.; *Strategy and Policy of Reindustrialisation for Innovative Development of Russia*; Institute for New Industrial Development (INID), 2014.

Bodrunov, S. D. Integration of Science and Education Production and New Industrialisation of Russia. *Vedomosti* (Published Together With the Wall Street Journal & FT) **2014**, 215 (3719), 17.

Bodrunov, S. D. Grinberg R. S.; Sorokin D. E.; ReIndustrialisation of the Russian Economy: Imperatives, Potential, Risks. *Econ. Reviv. Russ.* **2013**, *No. 1 (35)*, 19–49.

Bodrunov, S. D. The Transition to the New Industrial Society of the Second Generation: The General Cultural Dimension. *Econ. Reviv. Russ.* **2017**, *1* (51), 4–11.

Bodrunov, S. D. Technology Convergence a New Basis for Integration of Production, Science and Education. *Econ. Sci. Mod. Russ.* **2018**, *1*, 9–19.

Bodrunov, S. D. New Industrial Society: Structure and Content of Social Production, Economic Relations, Institutions. *Econ. Reviv. Russ.* **2015**, *4* (46), 9–23.

Bodrunov, S. D. Reindustrialisation of the Russian Economy Opportunities and Limitations. *Free Econ. Soc. Russ.* **2014**, *1*, 15–46.

Boyes, H.; Hallaq, B.; Cunningham, J.; Watson, T. The Industrial Internet of Things (Iiot): An Analysis Framework. *Comput. Ind.* **2018**, *101*, 1–12.

Burnham, J. *The Managerial Revolution: What is Happening in the World*; A John Day Book, 1941.

Bushueva, I. V. New Strategic Priorities of Tourism Development in Russia. *Service Plus* **2019**, *13*(4) 27–29.

Buzgalin, A. V. Kolganov, A. I. The Simulacrum Market: A View Through the Prism of Classical Political Economy. *Philos. Econ.* **2012**, *3*, 181–193.

Buzgalin, A. V.; Kolganov, A. I. ReIndustrialisation as Nostalgia? Polemical Notes On the Targets of Alternative Socioeconomic Strategy. *Sotsis* **2014**, *No. 1*, 80–94.

Buzgalin, A. V.; Kolganov, A. I. Simulacra Market: A View through the Prism of Classical Political Economy. *Alternatives* **2012**, *2*, 65–91.

Caragliu, A; Del Bo, C.; Smartness and European Urban Performance: Assessing the Local Impacts of Smart UrBan Attributes. *Innovation: Eur. J. Social Sci. Res.* **2012**, *25*(2), 97–113.

Castells, M. *Information Epoch: Economy, Societies and Culture*; Shkaratana, O. I., ed. and trans.; Higher School of Economics, 2000.

Chase, S. A. *New Deal*; The Macmillan Company, 1932.

China Sharing Economy Market to Exceed 9 Trln Yuan: Report. *Xinhua*, November 2, 2019. http://www. xinhuanet.com/English/2019-11/02/C_138523206.htm (accessed June 7, 2022.

Darkin, S. M. Strategic Problems of the Russian Far East. *Manage. Consult.* **2016**, *1*(3), 68–76.

Dementiev, V. E. *Long Waves of Economic Development and Financial Bubbles*; CEMI RAS, 2009.

Dementiev, V. E. Productivity Paradox in the Regional Dimension. *Reg. Econ.* **2019**, *T. 15*, *1*, 43–56.

Desai, R. *Geopolitical Economy: After US HeGemony, Globalization and Empire*; Pluto Press, 2013.

Desai R. *After American Hegemony, Globalization and Empire: Monograph*; Bodrunov, S. D., ed.; S. Yu. Witte, 2020.

Desyatnichenko, D. Y.; Desyatnichenko, O. Yu; Theoretical Aspects of Forming The Strategy of Development of Recreation and Tourism In the Region. *Manage. Consult.* **2016**, *4*, 150–157.

Drucker, P. *The Age of Discontinuity; Guidelines to Our Changing Society*; Harper and Row, 1969.

Eliseeva, I. I. Platonov, V. V.; Dynamic Potential the Missing Link in the Study of Innovation Activity. *Finance Bus.* **2014**, *4*, 102–110.

Ermakova, J. Technological Priorities as the Basis for Scientific and Technological Development of the Region's Industrial Complex. *Bull. Orenburg State Univ.* **2012**, *8* (144), 105–109.

Fersman, A. E. Geochemistry: In 4 Vol. T. 2. L., 1934.

Foresight "Russia": Design of a New Industrial Policy; *Proceedings of St. Petersburg Economic Congress (SPEK-2015)*, Bodrunov, S. D., ed.; Cultural Revolution, 2015.

Galbraith, J. K. *The Affluent Society*; Houghton Mifflin, 1958.

Galbraith, D. K. *The Return: A Monograph*; Bodrunov, S. D., ed.; Cultural Revolution, 2017.

Galbraith, D. K. *Society of Abundance* (Translation from English; Scientific Editor of the Russian Edition S.D. Bodrunov); Olimp-Business, 2018.

Galbraith, J. K. *The New Industrial State*; Princeton University Press, 1967.

Germany Trade & Invest. Industrie 4.0 – Germany Market Report and Outlook. Https://Www.Gtai.De/GTAI/ Content/EN/Invest/_Shareddocs/ Downloads/GTAI/Industry-Over Views/Industrie4.0-Germany-Market-Outlook-Progress-ReportEn.Pdf

Garipov, F. N.; Gizatullin, H. N. Stability of Functioning of Production and Economic Systems. *Reg. Econ.* **2012**, *4*, 116–122.

Giffinger R.; Gudrun H. Smart Cities Ranking: An Effective Instrument for the Positioning of Cities? *ACE: Archit. City Environ.* **2010**, *4*(12), 7–26.

Gizatullin, H. N. Garipov, F. N.; Garipova, Z. F.; Management Issues of Structural Transformation of the Regional Economy. *Reg. Econ.* **2018**, *14*(1), 43–52.

Glazyev, S. Yu. Prospects of Formation in the World of a New VI Technological Way of Life. *WORLD (Modernization. Innovation. Development)* **2010**, *1, 2*(2), 4–10.

Glazyev, S. Yu. On External and Internal Threats to Russia's Economic Security in the Context of American Aggression. *Manage. Bus. Admin.* **2015**, *1*, 4–20.

Govtvan, O. D. Gusev, M. S.; Ivanter, V. V.; Xenofontov, M. Y.; Kuvalin, D. B.; Moiseev, A. K.; Porfiriev, B. N.; Semikashev, V. V.; Uzyakov, M. N.; Shirov, A. A. System of Measures to Restore Economic Growth In Russia. *Prob. Forecast.* **2018**, *1*, 3–9.

Golovnin, M. Yu. Challenges for Russia's Economy from Global Financial Markets; *Proc. World Econ. Forum Russ.* **2019**, *218*(4), 314–319.

Greenberg, R. S. Smart Factories Need Smart People and Smart Economy. *Econ. Reviv. Russ.* **2016**, *4*(50), 154–157.

Guryev, A. I. *What Russians Didn't Like to Drive Fast? The Story of A Doomed Project*. OOO Izdat, KOSTA, 2009.

Harvey, D. *Rebel Cities: From the Right to the City to the Urban Revolution*; Verso, 2012.

Income Inequality and Economic Growth; Buzgalina, A.; Traub-Mertz, M.; Voeikov, M., Eds.; Cultural Revolution, 2014.

Integration of Production, Science and Education and Reindustrialisation of Russian Economy: Proceedings of the International Congress; Bodrunov, S. D., ed.; *Revival of Production, Science and Education in Russia: Challenges and Solutions*, LENAND, 2015.

Issberner, L.R.; Lena F. Anthropocene: Scientific Disputes, Real Threats. *UNESCO Courier* **2018**, No. 2. https://ru.unesco.org/courier/2018–2/ Antropocen-Nauchnye-Spory-Realnye-Ugrozy (accessed June 7, 2022).

Karlik, A. E.; Industry as a Structural Element of St. Petersburg Development. *St. Petersburg Offers* **2017**, *2*(25), 11–12.

Kleiner, G. B. Intellectual Economy of the New Century: Post-Knowledge Economy. *Econ. Reviv. Russ.* **2020**, *No. 1(63),* 35–42.

Kovalchuk, M. V. Convergence of Science and Technology A Breakthrough to the Future. *Russ. Nanotechnol.* **2011**, *6*(1–2) 1–32.

Kovalchuk, M. V.; Naraikin, O. S.; Yatsishina, E. B. Convergence of Sciences and Technologies and Formation of New Noo Sphere. *Russ. Nanotechnol.* **2011**, *6*(9–10), 10–13.

Krasilshchikov, V. A. *Follow-Up of the Last Century: Russia's Development in the Twentieth Century from the Perspective of World Modernization*; Russian State Library of Russia, 2010.

Krasilshchikov, V. A. Modernization and Russia At the Turn of the 21st Century; *Iss. Philos.* **1993**, 7, P. 40–56.

Kudrin, B. I. Researches of Technical Systems as Communities of Products—Technocenoses. System Research. Methodological Problems; Yearbook 1980; *Nauka* **1981**, 236–254.

Kudrin, B. I. *Technogenesis: Globalistics: Encyclopedia*; I. I. Mazur, I. I; Chumakov, A.N., eds.; Center for Research and Applications Dialogue; OJSC Publishing House, 2003.

Kuzyk, B. N.; How to Successfully Implement the Strategy of Innovation Development of Russia. *Mir Rossii* **2009**, *4*, 3–18.

Kuleshov, V. V.; Alexeev, A. V.; Yagolnitser, M. A. Assessing The Role of Natural Resources in a Country's Economic Growth: Cognitive Analysis and Decision-Making. Interexpo GEO-Siberia. XIV International Scientific Congress. April 23–27. *Novosibirsk* **2018**, *1*, 297–307.

Kuleshov, V. V.; Seliverstov, V. E.; Strategy for Socioeconomic Development of Siberia: Institutional Conditions and Implementation Mechanisms; *Econ. Reg.* **2005**, *4*, 5–23.

Kvint, V. L. To Setting Priorities. Budget, 2016. No.11. pp. 78–81.

Kvint, V. L. Strategy for the Global Market: Theory and Practical Applications New York: Routledge, **2015**. 548 p. ISBN 978-1-315-70931-4. DOI: 10.4324/9781315709314.

Kvint V. L. Strategic Economic Influence of a Global Negative Trend of Terrorism and Extremism. *Administrative Consulting.* **2016**, No. 6, pp. 14–25.

Kvint V. L., Bodrunov S. D. Strategizing Societal Transformation: Knowledge, Technologies, Noonomy. *St. Petersburg: S.Y. Witte INID.* **2021.** 351 p. ISBN 978-5-00020-083-4.

Kvint V. L., Okrepilov V.V. Quality of Life and Values in National Development Strategies. *Herald of the Russian Academy of Sciences.* **2014.** Vol. 84. No. 3. Pp. 188-200. DOI: 10.1134/S1019331614030058.

Kvint V. L. To the Analysis of the Formation of a Strategy as a Science // Herald of CEMI. 2018. Vol. 1. Issue 1. URL: https://cemi.jes.su/s111111110000121-6-1. DOI: 10.33276/S0000121-6-1.

Kvint, V. L. The Concept of Strategizing. Vol. I. – SPb.: NW M RANEPA, 2019. – 132 p. ISBN: 978-5-89781-628-6, 978-5-89781-629-3.

Kvint, V. L. The Concept of Strategizing: Monograph / V.L.Kvint. - 2nd Edition. - Kemerovo: *Kemerovo State University,* **2022.** – 172 p. ISBN 978-5-8353-2844-4. DOI: 10.21603/978-5-8353-2562-7

Kvint V. L. Development of Strategy: Scanning and Forecasting of External and Internal Environments. Administrative Consulting. 2015. No. 7. pp. 6–11.

Kvint V. L. The Global Emerging Market: Strategic Management and Economics. Moscow: *Business Atlas,* **2012.** 627 p. ISBN 978-5-9900421-6-2

Kvint V. L. Theoretical Basis and Methodology of Strategizing of the Private and Public Sectors of the Kuzbass Region as a Medial Subsystem of the National Economy. *Russ. J. Ind. Econ.* **2020.** Vol. 13. No. 3. pp. 290–299. DOI: 10.17073/2072-1633-2020-3-290-299.

Kvint, V. L. Strategizing: Theory and Practice. *Tashkent, Tasvir,* 2018. 160 p. ISBN 978-9943-4003-7-5 (Uzbek).

Kvint, V. L. Semi-trailer Territory. Does Russia Need a Spatial Development Strategy? *Ogoniok.* **2019.** No. 10. pp. 8–9.

Le Roy, E. *The Idealistic Requirement and the Fact of Evolution*; Boivin & Cie, 1927.

Leksin, V. N.; Porfiriev, B.N.; Assessment of the Effectiveness of State Programs of Socioeconomic Development of the Regions of Russia. *Probl. Prognostication* **2016**, *4*(157), 81–94.

Lobin, M. A. Strategy 2030. Position of the Union of Industrialists and Entrepreneurs of St. Petersburg. *St. Petersburg Offers* **2017**, *2*(25), 20–21.

Mahloop, F. *Knowledge Production and Dissemination in the United States*; Progress, 1966. (The Production and Distribution of Knowledge in the United States; Princeton, 1962).

Malakooti, B. *Operations and Production Systems with Multiple Objectives*; Wiley, 2013.
Masuda, Y. *The Information Society as Postindustrial Society*; World Future Soc., 1983.
Mechanic, A. We Can't Do Without Our Worms. *Expert* **2014**, *37*(914). https://expert. Ru/Expert/2014/37/Bez-Svoih-Chervyakov-Ne-Obojdemsya (accessed June 7, 2022).
Mechanic, A. Machines for Children and Grandchildren. *Stimulus* **2017**. https://stimul.online/Articles/Sreda/Stanki-Dlya-Detey-I-Vnukov (accessed June 7. 2022).
Medvedev, Yu. An Organism Whose DNA Contains 6 "Letters" Has Been Created. XXII Century: Discoveries, Expectations, Threats. *Popul. Sci. Portal* **2017**. https://22century.Ru/BiologyAnd-Biotechnology/42655 (accessed June 7, 2022).
Meijer, A. Bolívar, M. Governing The Smart City: A Review of the Literature on Smart Urban Governance. *Int. Rev. Admin. Sci.* **2016**, *82*(2), 392–408.
Mikulski, K. On Conceptual Elaboration of Tasks for Modernization of Russian Economy. *Soc. Econ.* **2010**, *12*, 5–12.
Minakir, P. A.; Regional Strategies and Imperial Ambitions. *Spatial Econ.* **2015**, *4*, 7–11.
Muratshina K. G. China-2050: Specifics of Strategy Formation. Izvestia of the Ural State University, Ser. 3. *Soc. Sci.* **2010**, *3*(80), 85–92.
Nekipelov, A. D.; Globalization and Strategy of Russia's Economic Development. *Probl. Forecast.* **2001**, *4*, 3–16.
Novikov, Yu. Y.; Rezhabek, B. Г.; Contribution of E. Le Roy and P. Teilhard De Chardin in Development of Noosphere Concept. *Probl. Reg. Ecol.* **2010**, *1,* 88–94.
Novikova, I. *The Russian Far East: Strategic Development of the Workforce*; Apple Academic Press, 2020.
Novikova, I. V. *Concept of Employment Strategy in Digital Economy*; Kemerovo State University, 2020.
Ohno, T. *Just-In-Time for Today and Tomorrow*; Productivity Press, 1988.
Okrepilov, V. V.; Quality of Life: Country Strategy 2030 Guidelines. *Quality Economy* [Online] **2017**, *No. 1 (17)*. http://eq-journal.ru/Pdf/%D0%9E%D0%BA%D1%80%D0%B5%D0%BF%D0%B8%D0%BB%D0%BE%D0%B2.Pdf (accessed June 7, 2022).

Osipenko, A. S.; Technological Transfer in the System of Innovative Development of Industry. *Econ. Reviv. Russ.* **2014**, *1*(39), 83–88.

Phelps, E. *Mass Prosperity: How Grassroots Innovation Became A Source of Jobs, New Opportunities and Change*; Gaidar Institute Publishing House; Liberal Mission Foundation, 2015.

Phelps, E. S. A Good Economy for China. *Project Syndicate,* June 17, 2016. https://www.project-syndicate.org/Commentary/China-Innovation-Good-Economy-By-Edmund-S-Phelps-2016–06 (accessed June 7, 2022).

Piketty, T. *Capital In the 21st Century*; Éditions Du Seuil, 2013.

Platonov, V. V.; Solow's Paradox Twenty Years Later, or on Investigating the Impact of Innovation in Information Technology On Productivity Growth. *Finance Bus.* **2007**, *3*, 28–38.

Polyakova, I. L.; Ermakova, J. Standardisation in the Regional Hospitality Industry: Directions, Main Stages. *Bull. Orenburg State Univ.* **2015**, *8*(183), 116–121.

Porfiriev, B. N.; Low-Carbon Development Paradigm and Strategy of Climate Change Risk Reduction for Economy. *Probl. Forecast.* **2019**, *2*, 3–13.

Pride, V.; Medvedev, D. A.; Phenomenon of NBIC-Conversion: Reality and Expectations. *Philos. Sci.* **2008**, *1*, 97–116.

Prosvirnov, A. A.; New Technological Revolution Sweeps Past Us. *Proatom Agency*, December 11, 2012. http://www.proatom.ru/Modules.Php?Name= News&File=Article&Sid=4189 (accessed June 7, 2022).

Razumeeva, V. Railways Are Gaining Speed. *Business Guide (Railway Transport) (Supplement to the Newspaper* Kommersant*)* **2009**, *239(4294)*. http://www.Kommersant.ru/Doc/1292160 (accessed June 7, 2022).

Roco M.; Bainbridge, W. Converging Technologies for Improving Human Performance: Nanotechnology, Biotechnology, Information Technology and Cognitive Science; Arlington, 2004.

Sadovnichaya, A. V. *Strategy of Exhibition and Fair Activity*; IPC NRU RANEPA, 2019.

Sakaya, T. Value Created By Knowledge, or the History of the Future. *The New Post-Industrial Wave in the West: An Anthology*, Inozemtseva, V. L, ed.; *Academia* 1999, 337–371.

Samir, A. *Russia and the Long Transition From Capitalism to Socialism*; Monthly Review Press, 2016.

Samir Amin. *October 1917 Revolution, a Century Later.* Daraja Press, 2017.

Schwab, K. *The Fourth Industrial Revolution*, E Ltd, trans. and ed.; 2017. https://www.litres.ru/Klaus-Shvab/Chetvertaya-Promyshlennaya-Revoluciya-21240265/Chitat-Onlayn (accessed June 7, 2022).

Schmidt, K. *Nomos of the Earth in the Law of Nations Jus Publicum Europaeum*; Vladimir Dal, 2008.

Shirov, A. Problems of Reproduction In the Modern Russian Economy. *Voprosy Politicheskoy Ekonomiki* **2019**, *2*, 37–46.

Shirov, A. Socioeconomic Forecast as a Mechanism of Strategic Economic Management. *Budget* [Online] **2019**, 1. http://bujet.ru/Article/364772.Php (accessed June 7, 2022).

Sen, A. *Development as Freedom*; Novoye Izdatelstvo, 2004.

Simic, S. Need, Not Greed. *The Guardian*, January 25, 2007. https://www.theguardian.com/commentisfree/2007/jan/25/post997 (accessed June 7, 2022).

Smith, K. *What is the 'Knowledge Economy'? Knowledge-intensive Industries and Distributed Knowledge Bases*; United Nations University, Institute for New Technologies, 2000.

Smotritskaya, I. I. New Economic Strategy Requires A New Quality of Public Administration. *Bull. Inst. Econ. RAS* **2017**, *5*, 7–22.

Stiglitz, D.; Sen, A.; Fitoussi, J. P. Misjudging Our Lives: Why Does GDP Not Make Sense? Report of the Commission on the Measurement of Economic Performance and Social Progress, Kushnareva, I., trans.; Drobyshevskaya, T., ed. Gaidar Institute Publishing House, 2015.

Stiglitz, J. *The Price of Inequality: How The Stratification of Society Threatens Our Future*; Eksmo, 2015.

Sorokin, D. E.; Sukharev, O. S. Structural and Investment Tasks of Russia's Economic Development. *Econ. Taxes Law* **2013**, *3*, 4–15.

Sorokin, D. E.; Conditions for Transition to an Innovative Type of Economic Growth. *MIR (Modernization. Innovation. Development): Scientific-Pract. J.* **2010**, *1*, 2(2), 26–36.

Spohrer, J.; NBICS (Nano-Bio-Info-Cogno-Socio) Convergence to Improve Human Performance: Opportunities and Challenges, Roco, M.; Bainbridge, W.; eds.; *Converging Technologies for Improving Human Performance: Nanotechnology, Biotechnology, Information Technology and Cognitive Science* [Online] **2004**, 225–235. http://www.wtec.org/Convergingtechnologies/Report/NBIC_Report.pdf

Sziebig, G.; Korondi, P.; Effect of Robot Revolution Initiative in Europe–Cooperation Possibilities for Japan and Europe. *Sciencedirect* **2015**, *48*(19), 160–165.

Tatarkin, A. I.; Sobering Up After Market Euphoria is Delayed, But It's Happening Nonetheless: Interview. *City* 812 **2014**, *32*, 21–23.

The Global Wage Report 2014/15: Wages and Income Inequality. International Labor Organisation, 2015.

The New Industrial Society: Origins, Reality, Future (Selected Materials of Seminars, Publications and Events of the Institute for the New Industrial Society); Bodrunov, S. D., ed.; S. Yu. Witte, 2017. .

The New Industrial Society: Origins, Reality, Future, Volume II. (Selected Materials of Seminars, Publications and Events of the Institute of New Industrial Development), Bodrunov, S. D., ed.; S. Yu. Witte, 2018.

The New Industrial Society: Origins, Reality, Future. Noonomy, Volume III. (Selected Materials of Seminars, Publications and Events of the Institute of New Industrial Development), Bodrunov, S.D., ed.; S. Yu. Witte, 2019.

The New Industrial Society: Origins, Reality, Future. Noonomy, Volume IV (Selected Materials of Seminars, Publications and Events of the Institute of New Industrial Development), Bodrunov, S.D., ed.;. Yu. Witte, 2020.

The Third World: Half A Century Later, Khoros, V. V. G., Malysheva, D. B., eds.; IMEMO RAS, 2013.

The Word on the Law and Grace, Platonov, O. A., ed.; Institute of Russian Civilization, 2011.

Tillema, S.; Steen, M.; Co-Existing Concepts of Management Control: The Containment of Tensions Due to the Implementation of Lean Production. *Manage. Account. Res.* **2015**, *27*, 23–27.

Tkachenko, S. S.; On the Development Strategy of Domestic Machine-Tool Industry Until 2030 from the Perspective of Blank Production. *Metall. Mech. Eng.* **2019**, *5*, 26.

Tolstoukhova, N.; The Machine Requires Benefits. *Rossiyskaya Gazeta*, October 28, 2018. 28.10.2018. https://rg.ru/2018/10/28/Minpromtorg-Projected-Growth-of-Both-Industries-V-Stankostroenii.html

Tretiak, V. P. *Regional Foresight: Possibilities of Application*; Znanie, 2012.

Tsatsulin, A. N.; Approaches to Economic Analysis of Complex Innovation Activity. *Proc. St. Petersburg State Econ. Univ.* **2013**, *2*(80), 12–21.

Understanding Human-Machine Networks: A Cross Disciplinary Survey, Tsvetkova, M.; Yasseri, T.; Eric, T.; Meyer, J.; Pickering, B.; et al., eds.; Cornell University Library [Online] **2015**. https://arxiv.org/abs/1511.05324 (accessed June 7, 2022).

Valentey, S.; Bakhtizin, A.; Kolchugina, A. Readiness of Regional Economies for Modernization. *Federalism* **2018**, *3*, 143–156.

Veblen, T. *The Engineers and the Price System*; Batoche Books, 2001.

Vernadsky V. I. *Scientific Thought as a Planetary Phenomenon*; Nauka, 1991.

Wadell, W.; Bodek, N. *The Rebirth of American Industry*; PCS Press, 2005.

Wolff, E. N. *Poverty and Income Distribution*; Wiley-Blackwell, 2008.

World Social Forum 2016. *Global Justice Now* 2016. https://www.globaljustice.org.uk/event/world-social-forum-2016 (accessed June 7, 2022).

Wright, E. O. Perrone, L. Marxist Class Categories and Income Inequality. *Amer. Sociol. Rev.* **1977**, *42*(1), 32–55.

Zubkova, E. Russian Machine Tool Industry: A Thorny Path to Success. All Industrial Regions of Russia [Online] **2017**, March 29. http://Www.Promreg.Ru/Articles/ RossiJskoe-Stankostroenie-Ternistyj-Put-K-Uspehu (accessed June 6, 2022).

Appendix

NEW WAYS AHEAD FOR NATIONAL ECONOMY (REGARDING POSSIBLE RUSSIA'S DEVELOPMENT STRATEGY)[1]

Russia has been relatively successful overall in dealing with the 2020 crisis caused by the pandemic and falling world prices, as well as oil production cuts under the OPEC agreement. Nonetheless Russia's GDP fell by 3.1% and industrial output by 2.9% in 2020, which is less than in majority developed European countries.[2] The unemployment rate, which stood at 6%, and mass bankruptcy of SMEs have been contained, while the incidence of COVID-19 in Russia is lower than in most European countries. Much has been done to help low-income families.

The government's economic and social support measures have created the preconditions for overcoming the negative effects of the crisis more quickly than in some other national economies.

This allows to expect Russia to achieve a slightly better position in the global economy and the international division of labor in the post-pandemic period.

At the same time, the 2020 crisis was another year of 7 years of stagnation, which began in 2014 under the impact of anti-Russian sanctions,

---

[1] Contribution from the Russian Free Economic Society was sent on 2/12/21 to the Government of Russia and was developed within the framework of the task set by the President of the Russian Federation before the public institutions and expert community of the country to constructively participate in the improvement of the strategy of national economic goals and the draft of strategy for socioeconomic development of the Russian Federation, which ensures achievement of national goals for the period until 2030. Document was prepared on the basis of materials and reports by experts of the Russian Academy of Sciences (RAS): Academician of RAS, Dr. of Economics B. N. Porfiriev, Academician of RAS, Dr. of Economics A. G. Aganbegyan, Academician of RAS, Dr. of Economics A. A. Dynkin, Academician of RAS, Dr. of Economics S. Yu. Glazyev, Academician of RAS, Dr. of Economics V. A. Kryukov, Academician of RAS, Dr. of Economics V. L. Makarov, Academician of RAS, Dr. of Economics A. D. Nekipelov, Corresponding Member of RAS, Dr. of Economics A. A. Shirov, a Foreign Member of RAS, Dr. of Economics V. L. Kvint, Corresponding Member of RAS, Dr. of Economics A. R. Bakhtizin, Dr. of Economics A. N. Klepach, Corresponding Member of RAS, Dr. of Economics M. Yu. Golovnin, Corresponding Member of RAS, Dr. of Economics S. D. Bodrunov, Dr. of Economics E. B. Lenchuk et al. Here it is given in abridged form (for the full materials of expert sessions and conferences of the Russian Economic Community on the subject (see the website of the Russian Free Economic Society http://www.veorus.ru/)..

[2] Evaluation by the Institute of Research and Expertise (VEB.RF).

falling oil prices and accumulated internal structural problems. Although the Russian economy rebounded slightly in 2018–2019, the overall 7 year average annual growth rate was 0.25%, with real incomes declining by 13% (or an average of almost 2% per year). The main drivers of socioeconomic development in Russia have weakened: the share of fixed capital investment in GDP has fallen to 17% (a 10% drop), the share of the "knowledge economy" to 14%, exports as a whole have fallen by 20% in the last 6 years, and real disposable income by 14%. However, according to preliminary estimates by Rosstat, the population fell by 500,000 last year, including due to the pandemic, and the country is at risk of further depopulation.

The government adopted a plan to restore production, employment, and income to pre-crisis levels.

However successful this may be, there is an urgent need for a qualitatively new model of development and renewal of the social structure of society.

Last year marked the 12th anniversary of adopting the "Concept for the Long-Term Socioeconomic Development of Russia until 2020." It set the goal of a transition to an innovative and socially oriented economy, with the knowledge and high-tech industries as the main driver, and a middle class (with incomes comparable to those in developed countries) comprising 35–40% of the total population. Over the years, this challenge has not been fully addressed. The global economic crisis of 2008–2009 and anti-Russian sanctions have played their part. Still, in many ways, this is also a result of the fact that development and transformation objectives have taken second place to the priority of current activities and the focus on maintaining financial stability at all costs. The current model of economic and macroeconomic (including credit and monetary) policy, as the Russian President has repeatedly pointed out, has run out of steam.

It should be borne in mind that the internal and external operating conditions of the Russian economy have now changed fundamentally. The new world economic order is being formed based on advanced technologies. The role of knowledge as a basic economic resource and the role of the individual as its bearer is rising sharply. The share of innovative and knowledge-intensive products and services is growing, depreciating capital expenditure and investment in ageing industries and traditional areas of economic activity. The center of global economic development is shifting to the East Asian region. New drivers and mechanisms for implementing geo-political interests of the world's leading economies are emerging.

The urgency of abandoning the current paradigm of the country's economic development and developing a *new generation economy* based on knowledge-intensive industry of the coming technological order, high labor productivity, and competitiveness becomes extremely important for Russia in these conditions, as does the task of building a society without extreme inequality with a large middle class based on a new social contract.

At the same time, there are no objective obstacles for Russia to achieve significantly higher GDP growth rates in the nearest future (3–5 years).

The pandemic has aggravated the *problem of finding a balance between the acceleration of economic growth and preservation of human life, development of human wealth* (**health, knowledge, standard and quality of life, and an eco-friendly human environment**). Russia will have to choose an independent path, a new path that combines development that involves reduction of income lagging behind the developed countries, with development of environmentally friendly technologies that provide high-quality air, water, effective waste recycling, high energy efficiency, and optimal use of different types of energy.

It seems necessary to build a new model of economic policy (with two levels) based on a package of five key areas (vectors) of strategic transformation, designed for 10–15 years until 2030–35 and beyond, some of which should be clearly separated from the strategic long-term objectives and launched in 2022–2024, with their strict linkage, coordination, and necessary resources.

The first is a new social model of development.

The goal of overcoming extreme poverty must be transformed into poverty alleviation and expansion of the middle class. According to Rosstat's estimates, the poor (those who live below the subsistence level or whose income is less than 42% of the country's median income) include 18.8 million people, or 12.8% of the population (experts estimate that around 15–16% of the population). The low-income (1–3 subsistence minimum) make up[3] about 54–56% of the population, the middle class (4–12 subsistence minimum) 27–29%.

---

[3] Evaluation by the Institute of Research and Expertise (VEB.RF).

The introduction of a nationwide standard of public sector services and labor remuneration, with a significant increase (1.3–1.6 times within 3–5 years) in the share of labor costs in the national product, is important to significantly improve the living conditions of the population, especially the poor and the middle class, an important effect and condition of which is also a general acceleration of economic growth. It would also reduce excessive interregional wage differentials in the public sector and cost about 1.5 trillion rubles over the first 3 years (about 0.3–0.4% of the GDP). In the long run, the dynamics of wages for doctors, teachers, and scientists should be oriented toward the level of developed countries, bridging the gap with them, and thereby curbing the brain drain. Higher wages in the public sector would also push up wages in the private sector, given the overall improvement in economic efficiency and human capital.

Poverty can be reduced by increasing child benefits, unemployment benefits and, more importantly, by improving social protection (through natural resource rents and fairer taxation), increasing employment rates, and boosting income growth, for example, in sectors such as agriculture, where wages are about half of the national average. Nowadays, loss of employment and low unemployment benefits (despite their temporary increase in the implementation of anti-crisis measures) make people take any job and go into the shadows. An increase in unemployment benefits would support people's standard of living and prevent them from falling into poverty, while higher wages would encourage them to look for work more persistently. Increasing the amount of allowance to 80% of the salary paid during the previous year (or 6 months) during the first 3 months of the job search (down to 60% during the 4–6 months of the search and 50% in the following months provided the average wage is not exceeded, e.g., 30–35,000 rubles) would cost the budget system about 1.5–2 trillion rubles over 4 years, which does not appear unattainable.

In population aging, it is also impossible to achieve welfare growth without ensuring decent living conditions for pensioners and providing them with opportunities for active, including working and activities. An important element of the "Silver Age Economy" is the overdue increase in pensions and the development of pension provision in a direction that would determine the long-term rules of pension formation.

The following are the main types of savings. The mechanisms may vary but in general the focus should be on closing the gap between pensions and average wages; the ratio of average pension to average wage should increase from 30% (at present) to at least 32–33% by 2025 and 35–40% by 2035. This would require about 1.5 trillion additional rubles as early as 2024 (compared with the scenario of freezing the relative gap between pensions and average wages).

Overall, these measures, combined with accelerated economic growth of 3–4% or more per annum, would make it possible to offset the decline in real incomes between 2014 and 2020 and expand the share of the middle class to a third of the population by 2025 and to half by 2030. However, this requires a significant modification of the budget rule to avoid increasing payroll taxes and a significant monetary policy adjustment to provide long-term investment loans to the real economy, as sustainable long-term welfare growth can only be achieved by increasing labor productivity and maintaining overall high overall growth rates.

This requires, on the one hand, as recognized in many government policy documents, a significant increase in the savings rate (up to 25–27% of the GDP). In the Russian economy, national spending exceeds national savings by at least 2–3% of the GDP. This creates an opportunity for a significant increase in investment from domestic resources, provided that the overall business investment climate improves and capital outflows are reduced, and that there is targeted issuance within the limits (as in the US, EU, Japan, and other countries) allowed by the size of our foreign reserves or, as a last resort, that part of government savings (budget resources, sovereign wealth fund, and foreign exchange reserves) is directed to development instead of the excessive accumulation of financial assets. At the same time, the state must increasingly act as an institution of social development and coordination, and not just as an instrument of macroeconomic stability. By recapitalizing development institutions alone to support investments worth about 0.3–0.5% of the GDP over 4 years, this would raise the growth rate of the economy by 0.3–0.4% points per annum and launch new priority projects that change the quality of the Russian economy.

Such a strategy will require a revitalized planning institution to combine the benefits of the market and planned management and mobilize resources to implement the plans. It is necessary to develop a comprehensive system of forecasting and indicative planning of the country's socioeconomic development following the legislation on strategic planning. Based on the specifics and patterns of the current stage of the STP and the achieved level of socioeconomic development, the forecasts and projections should be made for periods of 15–20 years or more, medium-term ones for periods up to 10 years, and specific short-term ones for 1–3 years; the latter, including characteristics of all main macroeconomic parameters and instruments of implementation for basic Russian state development programs (including national projects), could be the basis for the formation of a 3-year plan. The advisory and guiding nature of the indicative plan should be combined with a high level of directive power for the public administration and state-owned companies.

Therefore, it would be advisable to establish a special government body to create such a system for planning the development of the national economy and monitoring the implementation of the plans.

On the other hand, increasing productivity and maintaining an overall high growth rate require a scientific and technological upgrade of the Russian economy. The share of the high-tech economy should rise from 21.8% of the GDP in 2019 to 24–25% in 2030.

The second track is thus a transition from lagging behind to a scientific and technological breakthrough and taking the lead in global scientific and technological rivalry.

The economic leaders of the future are technological leaders. The scientific and technological breakthrough is the main direction of accelerated economic development. The transition to new technology contributes to this balance between achieving high economic growth and high standards of living. It should also be borne in mind that health care and the entire human reproduction complex, including habitat cleansing, are set to become the biggest new technology-based industries in the future, and already are from now on; they are capital intensive and costly, so these trends will be accompanied by growth in GDP and investment.

In this context, the Presidential Decree's objective of keeping the Russian Federation among the world's top 10 countries in terms of R&D in the emerging environment could, in fact, lead to a freezing of the current position, as Russia is already among the top 10 countries in terms of R&D spending at the PPP level. A more ambitious goal of becoming one of the top five global scientific and technological leaders by 2035 could be set as a target.

To achieve these goals, it is advisable to increase R&D expenditure by a factor of 1.5 by 2025 and a factor of 3 by 2030. The driver of scientific and technological breakthroughs is likely to be the leading public and private research centers, together with universities, but not universities per se. A network of consortia or clusters of applied science centers or national laboratories with universities, corporate science centers and academic institutions is needed. Russia's position in the sphere of scientific and technological leadership should be at least no lower than its overall position in the global economy.[4] Meanwhile, the practice in recent years (in particular, in the sphere of high technologies in the military-industrial complex, aircraft and shipbuilding, nuclear energy and fuel and energy complex, microbiology, and virology, etc.) shows obvious opportunities for Russian science and advanced segments of the domestic new-generation industry to achieve this objective.

The Grand Challenges outlined in the Science and Technology Development Strategy (STDS) need to be translated into a range of scientific and technological priorities for the country in fundamental and applied areas: scientific for fundamental research and technological for applied development, in which Russia can make a breakthrough and become a leader.

Within these priority areas, there is a need to launch 10–15 major projects within the framework of the STDS, where, among other things, it is advisable to pilot the proposed mechanisms to improve the efficiency of the domestic science and technology sector.

Along with supporting the development of digital and quantum technologies and artificial intelligence, it is important to implement a comprehensive science and technology initiative for the development of microbiological, genomic, and medical technologies. The new quality of medicine is not only an essential social component, but also a priority and promising area for scientific and technological development, where lagging behind is unacceptable.

To reestablish the function of identifying technology priorities, coordinating basic and applied research, and developing it into innovative

---

[4] Russia now ranks 9th in the world in terms of R&D spending (in terms of purchasing power parity), and thus is 11.8 times behind China and 13 times behind the USA. R&D expenditure relative to GDP (1%) has been stagnant for almost 12 years. Whereas in 2008, we were at about the same relative level as China, China has now increased spendings to 2.23% of GDP. In the USA, they are at 2.83% and in South Korea they are at 4.55%. The National Science Project has envisaged that Russia will retain its fifth-highest full-time equivalent number of researchers among the world's leading countries (according to the Organization for Economic Cooperation and Development) from 2018 to 2021. However, according to the OECD, already in 2018, the Republic of Korea overtook Russia in this indicator, thus shifting it to the 6th position in the ranking.

large-scale projects, it is advisable to support the establishment of a state commission (committee) on science and technology that is above departmental and private interests and can set objectives for the scientific community and monitor their implementation in cooperation with the RAS and other institutes.

The most important task is to technologically reequip the existing industries and create new capacities in the high-tech sectors, increasing their volume to 1.5 times by 2025 and 3–3.5 times by 2035, accompanied by an aggressive formation of modern and advanced transport and logistics, engineering, communications, and energy infrastructure. To finance such expenditures (including investment in fixed capital), it is necessary to mobilize funds in the amount of at least 5 trillion rubles annually, which could be achieved employing a low-interest investment loan; there are objective possibilities for that given the relatively low inflation achieved in Russia, in which the banking system, 73% of which is under government control, exceeds the country's GDP (currently 110 trillion rubles).

The third direction is environmentally oriented development and the creation of an economy of nature preservation.

The priority for Russia should not be forced reduction of its carbon footprint, although this remains a key objective, but rather integrated environmental management. Reducing air pollution, providing clean water, creating a recycling industry (mainly full cycle), and developing a new forestry industry could be priorities.

Along with the development of a mechanism for accounting for and control of harmful emissions, including greenhouse gases and carbon dioxide emissions, and the organization of internationally recognized trade in carbon units, a scientifically based, correct assessment of the potential of forest and wetland ecosystems, including their area, structure, regulating role in ensuring water (hydrological) balance, mitigation of regional climate, and carbon sequestration capacity, is significant for Russia. While progress has been made in reducing the long-term costs of renewables, they cannot provide the sustainable energy supply for the population and economy that is guaranteed by hydrocarbon (primarily gas) and nuclear power generation. Therefore, as the energy sector evolves toward diversification, the balanced structure of energy production and consumption in Russia will be different from that in Western Europe while maintaining a reliance on these traditional sources for the foreseeable future. This does not underestimate the need for a significant increase in efforts to improve energy efficiency in production and consumption and the electrification of different modes of transport. We need to unlock the potential of traditional Russian energy developments in the field of superconductivity and fuel cells. In the foreseeable future, the contribution of the fuel and energy sector to the technological development of the Russian economy will increase, while its contribution to the GDP and the country's budget will relatively decrease.

The fourth direction is a new model of spatial development aimed at the rise of middle Russia and a new turn to the East and the Arctic.

Such a model would require the creation of a new level of territorial and macro-regional governance.[5] Financing of macro-regions seems appropriate through supraregional development funds or by coordinating regional development corporations. In the next few years, most of Russia's regions could gradually be transferred to a normal system of financing self-sufficiency, self-financing, and self-management instead of the subsidized system, which is dying out. There is a need to give the regions (partly concentrating resources at macro-regional level) additional budget revenues by transferring part of the VAT and MET to the regions. Not only special subsidies for poor regions, but a comprehensive program (support mechanism), including differentiated standards for inter-budgetary allocations, education, and maintenance of health facilities, would be advisable.

Special programs are needed for the development of Russia's backward regions on its western borders (Pskov, Novgorod, and Smolensk Oblasts), which lag dramatically behind their European neighbors and Belarus in terms of living standards and development dynamics. A similar program should be developed for the revival of other regions in the Russian Non-Black Earth region.

Spatial development must be balanced, relying not only on agglomerations but also on various forms of settlement, including rural settlements, seeking to preserve and improve the quality of rural life.

The fifth track of change is related to the Eurasian challenge.

The reintegration of the Eurasian space and the creation of a Greater Eurasia require Russia's significant contribution: ideological, project, and

---

[5] The boundaries of macro-regions may not coincide with federal districts.

financial. The burden of leadership requires a cost that can be recouped economically and politically by gaining appropriate control over assets and investing in building a national managerial, scientific, educational, and media elite in neighboring countries.

Integration efforts can be pursued along the following lines:

- Working with the engineering, scientific, medical, and educational elite, where the influence of Russian scientific and professional traditions and contacts are strong. Russia can contribute to the human capital formation of partner countries to a greater extent than our neighbors, which are many times larger than Russia in terms of investments and financial assistance.
- Creating a structural fund or funds of the EAEU, which could support joint development projects that have a high integration effect and contribute to the development of a common Eurasian infrastructure.
- Financing projects in the field of development of industrial cooperation and effective distribution of production in the EAEU space.
- Joint programs, especially in education, science, technology, and health, based on the model of the Union State of Russia and Belarus.
- Expansion of the EAEU by granting observer status (with a possible transition to associate member status in the future, if the EAEU accepts one, and then to union member) to countries such as Uzbekistan, Mongolia and, in the longer term, Azerbaijan and Afghanistan.
- The development of the SCO from a predominantly economic security-oriented organization to an economic union (partnership), including Iran and others.

By creating an attractive development model for its citizens, Russia has every opportunity to become a new center of attraction for neighboring countries, realizing the possibility of balanced, sustainable development, ensuring the unity and harmonization of economic progress, growth in human wealth, and environmental conservation goals.

Index

A
Additive and distractive technologies, 7–8
Article 7(7), 150

B
Bayh-Dowell Act, 144
Burden on budget, 132–133

C
Cognitive technology
  digitalization, 9
  role of information and, 8–11

D
Deindustrialization
  consequences, 130
Development of Science and Technology
  state programme, 136
Digitalization, 9, 134
Directive planning, 149

E
Export-oriented import substitution, 119

F
Food production
  industry strategy, 168–169
  regional aspect, 169

G
Germany's best innovation cluster, 144
Global economy
  relationships of people, 57
  transformational processes
    changing ownership, 22–26
    increased environmental pressure, 31–34
    increased risks, 37–40
    interference with human nature risk, 34–37
    strategic milestones, 39–41
    technological capabilities growing, 26–31
    technology changes, 18–22
Global transformation of society
  civilizational risks, 42–45
  noneconomic mode, 54–59
  rationality, 49
  rejecting economic rationality, 45
  strategy focused, 59–60
Goal-setting and planning tools
  forecasting, 94–95
  foresight, 94–95
  holistic national strategy, concept, 90–94
  interrelation, 94–95
  mission and foundational components, 100–103
  rules of, 87–90
  stages of, 95–99
  targeted programs, 103–106
  three approaches of, 86–87

H
Hybrid technology, 7

I
Indicative-selective planning, 149
Industrial and technological revolution, 12
  industry 4.0 and smart factories, 15–17
  new industrial revolution, 17–18
  production knowledge intensity, 12–15
  social development strategy, 17–18
Industry strategy, 166
  food production, 168–169

J

Japan
  production, science, and education integration, 143

K

Kvint's methodology, 166

L

Law No. 172-FZ, 150
  Russian Federation subjects, 152

M

Market, 149
Modernization
  economic and institutional conditions, 135–138

N

National innovation system (NIS)
  insufficient performance, 136
  role, 136
National projects
  digital economy, 112
  high standards, 113
  pathfinders, 112
  weaknesses, 112
National Technology Initiative (NTI), 139, 140, 141
National Technology Initiative Platform, 140
NBICS technology convergence, 5–7
  features, 6
New normal
  innovation process continuity, 116–118
  technological shift, 115–116
NIS.2, 56, 59
  development, 44
NOO production
  individual growth, 69–72
  meeting, 62–66
  new production, 69–72
  strategic development goals, 72–73
  work and needs, evolution, 61–62
Noonomy, 47, 58
  definition, 46
  global transformation of society
  civilizational risks, 42–45
  noneconomic mode, 54–59
  rationality, 49
  rejecting economic rationality, 45
  strategy focused, 59–60
  presupposes, 48
  two steps, 54

P

Plano Brasil Maior, 119
Post-industrial, 130

Q

Quality enhancement, 172

R

Regional strategic planning, 157
  effectiveness, 159
  experiences, 159–166
Re-industrialization, 129
  active industrial policy, 131
  economic modernization, 134
  economic policy priorities, 131
  favorable economic environment, 131
  goal, 130
  post-industrial concepts, 130
  risks, 132
Republican Research Institute of Intellectual Property (RRIPI), 136
Russian Academy of Sciences (RAS), 108
Russian economy
  burden on budget, 132–133
  education integration, 143
  financing of science, 137
  lower competitiveness, 132
  modernization, 115
  national technology initiative, 138–143
  objective, 142
  production, 143
  reduction in efficiency, 132
  science, 143
  structure of participants, 138
  technological lagging, 133
Russian Federation, 163, 170
  machine tool industry, 173–174

Index

mass migration, 158
national machine tool revival, 175
staffing of industry, 176
strategic objectives, 174–175
territorial aspect of development, 157
tourism, 170
Russian Venture Corporation (RVC), 139

S

Sectoral strategy, 166
  shortcoming, 168
  tourism, 170
Siberia Economic Development Strategy, 155
Sixth technological mode, 2
  challenges of 21st century, 3–4
  individualization of production, 4
  industry's features, 2–3
  significance of challenge, 4
Smart City, 162
Social production
  coordinating economic activities ways, 149
Societal development, target
  economic individual, 73–74
  quality of life, 74–77
  strategic focus on, 77–78
Socioeconomic development, strategic goals, 107
  national projects
    digital economy, 112
    high standards, 113
    pathfinders, 112
    weaknesses, 112
  Russian Academy of Sciences (RAS), 108
  strategic targets, 107
    RAS Academician, 108–111
    raw materials producers, 108
    re-industrialization, 111
    unfold financialization, 109
Socioeconomic developmental goals and parameters, 61
  needs, 66–69
  NOO production
    individual growth, 69–72

    meeting, 62–66
    new production, 69–72
    strategic development goals, 72–73
    work and needs, evolution, 61–62
  societal development, target
    economic individual, 73–74
    quality of life, 74–77
    strategic focus on, 77–78
Solow paradox, 148
Spatial and territorial strategy, 171
Strategic planning, 151
  Kvint's methodology, 166
  planning as a necessary tool, 147–148
  problems and prospects in Russia, 149–152
  regional development, 153
  regional socioeconomic development, 152–159
  tourism, 170
Strategic targets, 107
  RAS Academician, 108–111
  raw materials producers, 108
  re-industrialization, 111
  unfold financialization, 109
Strategizing national development, 79
  and goals
    prerequisites and conditions for, 79–83
    structure and linkages, 83–86
  goal-setting and planning tools
    forecasting, 94–95
    foresight, 94–95
    holistic national strategy, concept, 90–94
    interrelation, 94–95
    mission and foundational components, 100–103
    rules of, 87–90
    stages of, 95–99
    targeted programs, 103–106
    three approaches of, 86–87
  interests
    prerequisites and conditions for, 79–83
    structure and linkages, 83–86
  priorities
    prerequisites and conditions for, 79–83
    structure and linkages, 83–86
Sustainable development planning, 158

T

Technological development strategy
  import substitution role, 118–120
  increased needs satisfaction, 120–125
  inequality, 125–128
  re-industrialization, 120
  synergy of meeting needs, 122
Territorialization, 159
Tourism
  quality enhancement, 172
  Russian Federation, 170
  spatial and territorial strategy, 171
  strategic planning, 170–171
Transformational processes
  changing ownership, 22–26
  increased environmental pressure, 31–34
  increased risks, 37–40
  interference with human nature risk, 34–37
  strategic milestones, 39–41
  technological capabilities growing, 26–31
  technology changes, 18–22

W

World Technology Assessment Center (WTEC), 5